MATH ANXIETY REDUCTION

Robert D. Hackworth

H&H Publishing Co., Inc.

H&H Publishing Company, Inc.
2165 Sunnydale Blvd., Suite N
Clearwater, Florida 33575
Phone (813) 447-0835

Cover: Cade's Cove, North Carolina
By William A. Borgschulze

ISBN   0-943202-15-9

Printing is the last number
10 9 8 7 6 5 4 3 2 1

i

PREFACE

This book is written with a great deal of humility toward the task and a tremendous respect for its prospective readers.

For the past sixteen years I have worked in a mathematics lab. This environment is intended for students with academic backgrounds and levels of achievement which are below those now expected in American higher education. I have believed myself to be very successful in that work when compared with others who have attempted this type of teaching challenge, but the day-by-day interactions I have with many students have been, at times, highly frustrating. The origin of that frustration is the fact that obviously capable students, many of them mature individuals with proved competencies, often seem unable to perform anywhere near their potential when faced with math situations.

Because I have never placed much faith in a biologically or genetically determined intelligence and/or math aptitude, I have sought other hypotheses for the failure of many of these students. Ten years ago, the most common reason given by students was a strong dislike of mathematics. Today, it is more frequently stated as a fear — a strong debilitating fear — that interferes with most attempts at communication, instruction, and learning.

Once I accepted a fear of mathematics as a real cause of student failure, I began to look very differently at the problems they faced. I improved my observation skills of students at work and read the psychology texts explaining the behaviors of my students. I became convinced that these students have experienced traumatic failure in situations involving mathematics. The result has been an inability to function productively in other mathematics learning situations. All attention is centered on the fear itself rather than the mathematics that is being taught. It's hard to learn anything when you are not able to pay attention.

This text is a direct result of my observations and study. The material includes the strategies I have devised for overcoming the fear we commonly call "math anxiety." It also includes the information that a student must acquire to go beyond the fear and begin, again, learning mathematics.

This book may have a variety of uses — some planned and some reported to me as highly successful. Originally, I intended the book to be used by the individual suffering from math anxiety. For that reason, the material is written so that one person, working on her/his own, could achieve its major benefits. Under best conditions, however, the individual will have a group which is working on the same material. The group will provide the support, encouragement, and shared experiences which are needed by most readers.

Another highly successful use of the book has occurred with professionals attempting to help others who suffer from math anxiety. These professionals (teachers, counselors, learning center personnel, etc.) often are in need of the information contained in these pages. The psychological origins of anxiety, theories of learning, and necessary mathematics study skills are, sadly enough, not common elements of our professional training. Those areas of knowledge are today necessities for the proper teaching of all students and certainly for those suffering from anxiety. Hence this text has found its place in the professional libraries of many individuals and institutions. Also, it has served as the centerpiece for some staff development workshops. No doubt, there are many other practical uses in the original training or upgrading of all personnel who must assist others to deal successfully with mathematics.

I have been aided by some very special, sensitive people in the development of these materials and they deserve special mention in this preface. Jean Smith of Middlesex Community College, John Roueche of the University of Texas, Gladdys Church of the State University of New York at Brockport, and Claire Weinstein of the University of Texas have played significant roles in my own understanding of the content of this book. Without those special people this book could not exist.

Most importantly I want to thank all the students who have bravely faced their anxiety and tried a math course again. I wrote earlier, in the first paragraph of this preface, that I had a tremendous respect for such students. The more I understand about the anxiety, the greater my respect for these students' courage in again facing the situations that have been so painful in the past. It is my earnest hope that these materials will provide the information, attitudes, and new skills that will eliminate or greatly reduce the anxiety that has stifled the hopes and legitimate academic goals of many students.

Robert D. Hackworth
St. Petersburg Junior College
Clearwater, Florida 1985

TABLE OF CONTENTS

Chapter 1

Debunking Myths about Math Anxiety

Before beginning the instruction of this book, some of the basic understandings and beliefs of the material need to be stated and studied.

One of the major problems that make overcoming "math anxiety" a difficult task is the existence in our culture of deeply held beliefs about the nature of intelligence and, more specifically, the origins of the learning problems which some students encounter in mathematics.

The predominant belief systems of our culture play a major role in making students who fail feel weak, stupid, and incompetent. The learned belief systems of our culture provide rational, understandable reasons for student failure which accept the fact that strong, intelligent, competent students may fail mathematics despite their admirable qualities.

In this first chapter of the book, we're going to look at these belief systems that operate within the American culture (and others). Our objective is to make clear the wide difference of opinions that exists and, hopefully, cast some doubts upon the validity of those belief systems which contribute to the "math anxiety" problem.

TWO EXTREMIST VIEWS OF THE NATURE OF MATH ANXIETY

There are many different viewpoints on the problem denoted by "math anxiety." We will outline the two most extreme positions with the understanding that most people will hold belief systems that are somewhere between those two extremes.

At one extreme position, there are those who believe that:

a) Each individual is born with a certain intelligence which may differ greatly from the intelligence of other individuals. This native intelligence will include natural abilities to understand and handle mathematical concepts and skills, and

b) Those individuals who suffer a natural-biological shortcoming in intelligence experience failure as a natural consequence of those built-in deficiencies.

Myth: The brain is compartmentalized into school subject areas.

The viewpoint described above in its most stark and harsh terms considers the concept of "math anxiety" as fantasy and its name as a misnomer. "Math Anxiety" for those who hold this viewpoint is a failure to accept reality.

An equally extreme, but opposite, position on "math anxiety" is held by others who believe that:

a) Nearly all (98%) individuals possess the intellectual capability for understanding and handling mathematics concepts and skills, and

b) Those individuals who suffer "math anxiety" have had experiences which negatively affect their ability to be successful with mathematics.

Language

Art

This second viewpoint describes the position of this book; it generally fits the beliefs of those who originated the idea of "math anxiety" and have promoted strategies for overcoming the problem.

Myth: Some people are born with no mathematics compartment

WHO SUFFERS FROM MATH ANXIETY?

Regardless of one's viewpoint on the nature of math anxiety, there is general agreement that large numbers of our population openly express a fear and/or dislike of mathematics. It is estimated that at least 50% of the adult American population chooses to avoid, whenever possible, activities which require mathematics computations and/or thinking. The number of people affected by math anxiety is, by itself, enough to make the problem serious, but two factors facing students today further add to the difficulty. They are:

1) Options for jobs, college majors, etc. are greatly decreased when mathematics is avoided. For example, approximately 80% of the majors for a college student require more than the minimum mathematics courses.

2) The computer revolution is soon expected to separate our society into computer literates and computer illiterates. It is unlikely that math anxiety sufferers will be computer literate.

MATH ANXIETY AND THE WOMEN'S MOVEMENT

Some critics of math anxiety see a direct relationship between it and the women's movement. Although no cause and effect relationship exists,
there is no doubt that the women's movement has
questioned some traditional beliefs of our society
and encouraged the investigation of new causes for
the failure of many women in mathematics.

Certainly there is a correlation between partici-
pation in the women's movement and membership in
a math anxiety group. Females, in far greater
numbers than males, have sought help with their
math anxiety problems. This fact does not, however, prove that math anxiety is a women's movement problem. Nevertheless, it would not be surprising if Phyllis Schafly were to attack math
anxiety as a conspiracy designed to promote the
political interests of radical women's groups, but
a more reasonable explanation seems to hinge upon
a psychological understanding of anxiety in combination with the expectations placed on women
in the American culture.

There is, in the American culture, a general expectation that females will be more fearful and
more passive than males. This expectation may
explain the willingness of
more females to admit their
anxiety and join groups to
try to overcome the difficulty.
In that regard, the expectation plays a helpful role because admitting one's anxiety
is a necessary first step
toward working on the problem.
However, another aspect of
the cultural expectation for
females is that the expected
passive behavior may play a
significant role in pre-conditioning women toward negative reactions to anxiety
producing situations.

*"I'm not afraid of math,
but I sure don't like it."*

Males within the American culture are generally expected to be more courageous and active (aggressive) than their female counterparts. Males can be expected to be less likely to admit to their anxieties and, therefore, less likely to join groups to work on the problem. At the same time, the expectation that males should be active (aggressive) may have better protected many men in anxiety producing situations.

This book takes the position that large numbers of both women and men experience math anxiety. Cultural expectations may have contributed or detracted from the degree to which women or men suffer from anxiety, but both women and men frequently display behaviors that are counterproductive when interacting with mathematics.

Sex roles aside, math anxiety affects a surprising cross-section of the population. Math anxiety is not limited to unsuccessful people who never completed arithmetic. Doctors, lawyers, teachers, and business executives often admit (sometimes proclaim almost proudly) to math anxiety even though they have all survived education systems and vocations which impose some mathematics requirements. More surprisingly, there are accountants and even mathematics teachers who experience incompetence when faced with some tasks that would seem easy for anyone with their occupations. For example, there is a successful accountant who never balances his own personal checking account; he relies completely on the bank statement. As another example, there is a chairperson of a mathematics department who never teaches word problems because, as she candidly admits, she "can't figure them out."

The answer to the question:

"Who suffers from math anxiety?"

is at least 50% of the adult population, many of them in surprising occupations and positions.

WHAT ARE THE CHANCES FOR A MATH ANXIETIES CURE?

Some math anxiety proponents and/or programs are quite optimistic about the likelihood of curing someone who is suffering from the problem. This book wants to be very positive in its approach, but also realistic enough that readers will not expect amazing changes to occur without a major investment of time and, perhaps, the confrontation of painful memories and tasks. The individual who suffers from math anxiety almost always has at least a ten-year history with the problem. It was a long time growing or festering within the individual and it will take both time and effort to overcome. There is no easy, quick cure.

With the foregoing understandings, the claim can be made that almost everyone who now sees mathematics as a huge, impossible barrier to her/his other goals can expect to find ways through, over, and/or around that barrier. A complete cure may not be possible, but an ability to cope and survive is achievable by almost anyone willing and able to pick up this book and begin working on the problem.

WHO TEACHES THE MATH ANXIETY THAT IS LEARNED?

The belief behind this book is that math anxiety is learned. Nobody is born math anxious. Nobody is born with limited math ability. Nobody has a brain that is separated into compartments of school subject areas, and consequently, nobody has a math compartment that is empty.

Math anxiety is learned. If we were able to accurately study all the life experiences of a math anxious person, we would be able to discover the circumstances under which the math anxiety was learned.

Math anxiety is learned and there is always some significant "other person" in the learning situation that teaches the anxiety. There are two categories of teachers of math anxiety. There are "villains" and "loved ones."

Villains: In some cases, the teacher of math anxiety is an individual who behaves meanly and/or cruelly. These teachers of math anxiety are recognized and disliked by the learner, but nevertheless have their negative affect. Sometimes these "villains" are not instructors. Classmates, siblings, neighbors, and even parents are often the ones who teach math anxiety using mean or cruel behavior.

"Hey, I'm doing great in math now."

Loved Ones: In some cases, the teacher of math anxiety is perceived as acting kindly or lovingly. These teachers of math anxiety are loved and/or respected by the learner, but nevertheless have their negative effect. Sometimes these "loved ones" are not instructors, but, again, they may be classmates, siblings, neighbors, or parents.

Whether a teacher of math anxiety is a "villain" or "loved one" his/her behavior is almost always accompanied by the belief that "this will be good for the individual."

One of the objectives of this book is to make each reader aware of her/his own potential for teaching math anxiety. A knowledge of how math anxiety is taught will make it possible for you, the reader, to avoid passing on the type of behavior that creates and/or maintains math anxiety.

WHERE IS MATH ANXIETY LEARNED?

Most sufferers of math anxiety who clearly remember how their anxiety was created report that the situation occurred in school. The majority of these school-anxieties were created in elementary schools and many are associated with fractions.

Relatively few believe their math anxiety began in junior high school, but many youngsters who entered high school mathematics with a confidence based on successful earlier experiences encountered failure and frustration with beginning algebra or geometry.

A third and last major stumbling block in the math curriculum is a first course in calculus. Some interesting theories could be developed on the reasons why these three points in the math curriculum are the most common times for the creation of math anxiety. Perhaps there are unusually difficult content materials at these levels which could be eased with different course outlines. A more likely prospect, from this author's view, would consider the attitudes of the teachers at these three points in the curriculum. Elementary school teachers, especially when approaching the teaching of fractions, often express a fear or dislike of mathematics. Secondary teachers of algebra and geometry frequently see their subjects as being of a far higher status than the preceding elementary school arithmetic. Calculus teachers, too, often treat their subject with a reverence and an awe that communicates a belief that calculus can only be achieved by exceptional minds. A change in the attitudes and/or behaviors of mathematics teachers at these three points in the school curriculum may have a beneficial affect on the number and severity of math anxiety cases.

Some math anxiety is reported as created within the home, but relatively few sufferers remember their first fear occurring there. However, home environments may play key roles in pre-conditioning individuals for the negative experiences they first remember as occurring in schools.

Once math anxiety is experienced, both the home and school environments frequently are seen as maintaining, and sometimes strengthening, the fear. Rather than offering a safe environment in which math anxiety may be reduced over time, schools and homes are seen as threatening places by many anxiety sufferers.

HOW IS MATH ANXIETY REDUCED?

There are five essential steps in treating and/or reducing math anxiety. They are:

1) Developing an understanding of how anxiety is created and applying that understanding to the unique experiences of the anxiety sufferer.
2) Becoming acquainted with learning theory and the qualities of excellent instruction.
3) Acquiring the four necessary learning strategies for studying mathematics.
4) Finding methods of decreasing the counterproductive reactions to anxiety.
5) Experiencing a series of successful mathematics learning tasks.

Each of these steps is explained completely in the following chapters of this book, but the explanation is not sufficient for overcoming the anxiety. Each step requires understanding, introspection, and practice before the desirable objective can be achievable.

The first step is a combination of acquiring information and applying that information to your individual, unique experience. The step involves learning the way in which anxiety is created and then coming to an understanding of the particular situation in which your anxiety originated. An objective knowledge of the circumstances involved in creating your anxiety will not automatically eliminate the fear, but it will serve as the basis for the development of a new belief system about yourself and your potential.

"If I get through these math requirements, I will graduate."

The second step is defensive in nature; it is designed to protect the individual from regressing during the treatment. Most individuals seeking help with math anxiety are enrolled or planning to enroll in a math course. Because a math course is a particularly sensitive environment for an individual with math anxiety, the second step in treatment provides information on learning and instruction that will make the environment a positive one.

The third step of the treatment process identifies four areas of mathematics learning strategies and emphasizes the necessity for incorporating all four skill areas into a system of effective study skills.

The fourth element of this treatment process desensitizes the individual. Math anxiety greatly diminishes an individual's achievement because it arouses so much fear and limits the amount of attention that is even given to the mathematics. The desensitizing process develops strategies for calming the individual before the fear dominates the situation.

The fifth and final step in the treatment is a positive math learning experience. Anxiety sufferers have normally progressed past the point where their math anxiety was created and no reduction can occur until that point is revisited and re-experienced successfully. Repeated success in math learning tasks is necessary as part of the cure. This book provides the opportunity for repeated success in learning mathematics.

HOW TO USE THIS BOOK SUCCESSFULLY

Two important suggestions for using this book are:

1) Be active. From this point on, this book is interactive. You are expected to answer questions, discuss the material with others, observe yourself and others, and try to avoid behaviors that are self-defeating. Passivity is your enemy because it places you under the control of situations or other people. Actively work toward solving your problems with confidence that you can control the situations that you encounter. Maybe you don't always feel positive, but act the way you believe a positive thinking person would.

2) An individual can read and react to this book, but your work will be more effective if you can find a group to join; engage in full discussion with others. You need both the perceptions of others and their support. Knowing you are not alone in your suffering or your search for a solution will be most helpful when the doubts that you have lived with threaten to overcome your positive approach.

CHAPTER 1   LEARNING ACTIVITIES

1. Select two adults (parents, relatives, neighbors, etc.) that
    you remember from your childhood. Answer the following
    questions from their points-of-view.

    a. Is intelligence primarily determined by biology
        and/or genetics?

    b. Are males superior intellectually (mathematic-
        ally) to females?

2. Name three other people who seem to have math anxiety.
    Try to list their behaviors or symptoms.

3. Write a short story involving a kind, loving person who con-
    vinces another person that they are incapable of learning
    mathematics. The story may be fictional or real.

4. Write a short story involving a mean, cruel person who con-
    vinces another person that they are incapable of learning
    mathematics. The story may be real or fictional.

5. A complete cure for math anxiety is not likely, but a sub-
    stantial reduction is possible if the individual is able
    and willing to confront painful memories. This confronta-
    tion will require attempts to learn new behaviors. Are
    you willing to make the necessary commitment that will make
    a near-complete cure possible?

Chapter 2

Anxiety Is Learned

Sadly, but truly, our culture operates on many ideas that have little support or justification in terms of today's knowledge. Some such myths may have some social value in aiming toward a new, better society, but others actually limit the ability of many individuals to reach their potential and, hence, to make their possible contributions to a better society.

One of the worst myths that permeates American culture blames an individual's biology or intelligence for the anxiety that individual experiences. The truth is that **anxiety is learned** and few, if any, individuals go through a normal day's activities without experiencing some anxiety. This is because **anxiety is a normal human reaction** to any situation the individual feels is threatening. Sometimes that anxiety is low level and, according to many authorities, may have no serious negative effect on the individual's behavior or performance. At other times, the anxiety reaches a level where everyone agrees there is serious interference with the individual's capability for acting physically and/or intellectually.

"When I read carefully, work the problem, and check my answer, the course goes smoothly."

In this chapter of the book, the material presents a psychological experiment that shows clearly how anxiety is learned and, once learned, how it drastically affects the ability of the individual to react in her/his own best interest. The presentation is in an interactive format which encourages the reader to participate in developing a thorough understanding of the material.

That interactive format also urges the anxiety sufferer to **become active** - rather than passive - in working to overcome the anxiety. As you shall see, being active is a necessary condition for relieving anxiety.

DIRECTIONS FOR USING THE INTERACTIVE FORMAT

The interactive format you are going to use contains three key elements you must consciously and conscientiously use. These three elements are:

1 An introductory sentence or sentences which impart some information you are to deal with.
2 A question about that information.
3 The desired answer for that question.

The first two elements of the format will be indicated by a numbered paragraph. Immediately following each question the answer is given. For you to achieve the best possible understanding of the material, you should answer each question before looking at the one given by the book. Furthermore, it is desirable that you write each answer and, therefore, space is provided in the book for you for that purpose. The reason you are encouraged to write your answers is that the more senses you involve in a learning experience, the more you are likely to remember it.

1

Some people believe that anxiety is inherited, but others believe that anxiety is learned. What does this book's author believe? _____

Anxiety is learned.

2

Most people being interviewed for a job experience some anxiety. Is it possible that a job-seeker will have a heightened awareness that actually contributes to making a good impression? _____

Yes

3

Some people, during a job interview, feel their hearts pounding and have sweaty palms. Is it possible that a job-seeker will suffer enough anxiety that the interview will go poorly? _____

Yes

4

The same or similar experience may cause anxiety in one person, but not in another. Does this prove that some individuals were born to be anxious? _____

No. Each individual is preconditioned by her/his past experiences.

5

To study scientifically how anxiety is learned, human subjects are not used because of the possibility of long-term harm. Do you believe that psychological experiments inducing anxiety could create serious problems for their human being subjects? _____

It is hoped that your answer is "Yes." Experiments on human beings need to be carefully analyzed for their possible negative effects.

6

Some people are born and raised through infancy
with few traumatic events. Other people are
born and raised in environments which are clearly
hostile to their physical and emotional well-being.
Do all people reach the age of five-years-old
with equivalent backgrounds and histories?

_____

No. Each five-
year-old child has
a unique history
that has shaped
her/his life
and behavior.

7

An individual's history is unique. Some people
have backgrounds which have prepared them against
the negative effects of some circumstances. Other
people have backgrounds which may have actually
weakened them when placed in psychologically
dangerous circumstances. If two people encounter
exactly an identical situation, can they be expected to behave identically? _____

No. Each person
will behave in a
manner determined
by her/his unique
history.

8

An individual's unique history is pre-conditioning
her/his reaction to any new experience. Is it
possible for two different persons to have identical pre-conditioning? _____

No

9

Is it possible, even if not desirable, to put two
adults in identical anxiety-producing situations
and expect them to behave identically? _____

No. Their pre-
conditioning is as
important as the
new situation.

10

Human beings cannot be used in scientific research into anxiety because of the possible long-term negative effect and also because their pre-conditioning will distort any results. Might animals be used as the subjects? _____

> Yes, although some people also question the morality of this.

11

How might two dogs be chosen for a scientific experiment so that the effects of pre-conditioning would be minimized?

_____

Choose pups from the same litter.

12

Two pups from the same litter were chosen for a scientific experiment. Care was taken to choose two dogs that had similar life experiences. Is there any reason that the dogs should react in drastically different ways to similar situations?

_____

No. The effects of pre-conditioning were limited by the dogs' ages.

13

The first dog was named Alpha. Alpha was placed in a glass-sided box. Would the experimenters be able to observe Alpha clearly? _____

Yes

14

Alpha evidenced no resistance to being placed in the box. He began to investigate the container. Does Alpha's behavior indicate any anxiety about being in the box? _____

No

15
The floor of the box was wired so that a painful,
non-dangerous shock could be administered to its
occupant. The shock would not physically limit
the dog's behavior. With the electricity turned
on could the dog jump, howl, or urinate?

_____   Yes

16
A second feature of the box
was that one side was a hinged
door. That side would pop
open if it were pushed. Is
there a way out of the box
for Alpha? _____

         Yes

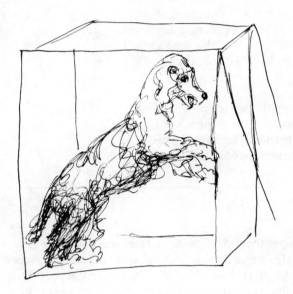

17
As Alpha inspected his cage,
the electricity was turned
on. Alpha urinated. He
jumped in the air. He barked.
He whined. Did urinating,
jumping, barking, or whining
get Alpha out of the box?

_____

         No

18
Alpha began running around the cage. As he ran,
he bumped (probably coincidentally) into the sides
of the cage. When he bumped the hinged side, it
popped open. Did running around the cage and
bumping the walls help Alpha escape from the cage?

_____   Yes

19
Seven seconds elapsed between the time the electricity was turned on and Alpha escaped. Were
those seven seconds painful for Alpha? _____   Yes

NEW IDEA

20

During those seven seconds when the electricity was on, Alpha was an **active** learner. He urinated, he jumped, he barked, he whined, and he ran in circles. Despite his pain, did Alpha **passively** accept the situation? _____

No

21

After Alpha escaped the cage, the experimenters turned off the electricity, caught the dog, and returned him to the box. What do you believe the dog's reaction was upon being returned to the box? _____

The dog resisted being returned to the box.

NEW IDEA

22

Alpha behaved differently when placed in the box for a second time. What was the difference in Alpha's behavior?

_____

The first time Alpha did not resist. The second time Alpha did resist.

23

Why did Alpha's behavior change the second time he was placed in the box? What do you think had occurred?

_____

Alpha seems to have learned that the cage is a hostile environment.

24

Alpha was in the box for the second time. When the electricity was turned on, Alpha didn't urinate, bark, jump, or whine. He immediately began running around and bumping the walls. Does it seem that Alpha had learned to discriminate between helpful and non-helpful behaviors? _____

Yes

25

On Alpha's second experiment in the cage, only two seconds elapsed between the turning on of the electricity and Alpha's escape. Were those two seconds physically painful for Alpha? _____

Yes

26

The experimenters returned Alpha to the cage for a third time. Do you believe that Alpha returned willingly? _____

No

27

When Alpha was placed in the cage for the third time, he went immediately to the hinged side, bumped it, and escaped. He didn't urinate, bark, jump, whine, or run in circles. He also didn't wait for the electricity to be turned on. Did Alpha experience any physical pain on his third visit to the box? _____

No

28

The experimenters returned Alpha to the box for a fourth time, but the dog immediately bumped the hinged side and escaped. Does it seem that Alpha had learned how to escape? _____

Yes

29

The experimenters continued to return Alpha to the box. The dog continued to escape. Had Alpha learned to act "intelligently" in the box environment? _____

Yes

NEW IDEA

30

As the experimenters continued to return Alpha to the box, the dog resisted less. Can you explain why the dog no longer considered the box a hostile environment?

_____

Alpha knew how to escape and this reduced the anxiety.

22

NEW IDEA

31
Does the information about this experiment indicate that Alpha is a "smart" dog? _____

No. There is no evidence here of the dog's intelligence.

32
Urinating and barking were two behaviors Alpha tried that were later discarded because they were not needed to escape. Can you name two other behaviors that Alpha discarded as unnecessary?

_____   _____

Jumping, whining, and running in circles.

33
Is it likely that Alpha would have "figured out" a way to escape from the box without trying other behaviors that were unnecessary? _____

No. Running in circles was an unnecessary behavior for escaping, but it was probably crucial to finding the successful escape strategy.

NEW IDEA

34
Alpha's successful discovery of an escape strategy contains a good lesson for anyone involved in a painful experience. Did Alpha passively accept his painful situation? _____

No. Be **active** in any learning situation.

35
Did Alpha's escape from the box occur because he was nice, quiet, responsible, mature, or thoughtful? _____

No. Those may be qualities of "good manners," but they are not the observable behaviors of "good learners."

NEW IDEA

36
Did Alpha's escape from the box occur
because he repeated the same unsuccessful
behaviors over and over again? _____

No. Repeating actions which
failed before is not likely
to lead to success.

37
Did Alpha's successful escape occur because
he tried different behaviors until he
discovered one that worked? _____

Yes

38
If you ever become the subject in an
anxiety-producing experiment what should
you do?

_____

Be active. Keep trying
different behaviors and
observe the results.

39
Can a successful learner experience pain
during the learning process? _____

Yes, Alpha did.

SOME IMPORTANT LESSONS FROM ALPHA

The experiment with Alpha provides some
valuable clues for behavior that results
from painful situations.
1   Alpha emerged from the experiment
     with some new skills.
2   Alpha learned new skills by acting
     very naturally in the situation.
3   Alpha was active physically while
     learning.
The next section describes another experiment, slightly changed, on a different dog.
The lessons to be learned from this second
experiment are also different from Alpha's
lessons.

40
The second dog used in the anxiety experiments was
named Baker. Baker was a brother of Alpha. Was
it expected that Baker had the same or similar pre-
conditioning as Alpha? _____

Yes

41
Baker was placed in the same
cage as the one used with
Alpha. Baker showed no resistance and began inspecting
his new environment. Was
Baker's behavior similar to
Alpha's first entry to the
box? _____

Yes

42
Although the cage used was the
same, the hinged side was now
locked shut for Baker's experiment. Was there any way out
of the box for Baker? _____

No

43
When the electricity was first turned on, Baker
urinated, barked, jumped, and whined. Was Baker
behaving similar to the way Alpha behaved? _____

Yes

44
By the time seven seconds had elapsed, Baker had
urinated, barked, jumped, whined, ran in circles,
and bumped the sides of the cage. Was Baker's
behavior similar to Alpha's? _____

Yes

45
Baker kept trying to escape from the hostile environment by trying different behaviors which were
observed by the experimenters. Were any of these
behaviors going to allow Baker to escape? _____

No

NEW IDEA

46
After twenty five minutes of actively trying to escape the electric shock, Baker laid down on the floor of the cage and did not move. Baker had changed his behavior. What was the change?

_____

Baker was now a passive rather than an active learner.

47
As Baker laid on the floor of his cage, the experimenters unlocked the hinged side and opened it. Baker saw the open side, but made no effort to escape. Was Baker physically unable to move? _____

No

NEW IDEA

48
Baker's refusal to escape from the electric shock was not caused by any physical failure. The barrier to his escape was psychological. Had Baker lost his psychological ability to actively seek relief? _____

Yes

ᴨᴨᴨᴨᴨᴨ   ᴨᴨᴨᴨᴨᴨ   ᴨᴨᴨᴨᴨᴨ   ᴨᴨᴨᴨᴨᴨ   ᴨᴨᴨᴨᴨ

SOME IMPORTANT LESSONS FROM BAKER

The experiment with Baker provides some valuable information about behavior that results from such situations.

1. Baker and Alpha were very much alike. The differences in their final behaviors is attributable to the differences in their experiments.
2. Baker behaved very naturally in the situation, but there was no way to escape. The lack of escape was not a failure of Baker.
3. The significant turning point for Baker's ability to cope with the situation came when he became physically passive.

49

One of the problems for an anxiety sufferer is the fact that physically there appears to be no handicap. Did Baker look quite capable of escaping the cage when the side was finally opened? _____

Yes

NEW IDEA

50

Extreme anxiety limits the sufferer from acting in his/her own best self interest. Did Baker have the ability to walk away from the cage after his twenty-five minute ordeal? _____

No. Physically the body was unharmed, but psychologically Baker lost his ability to respond.

NEW IDEA

51

To an **uninformed** observer might Baker's failure to escape be interpreted as a sign that the dog was stupid? _____

Yes. Any observer who did not know what had happened to Baker might make erroneous judgements about the dog.

52

Does the information about this experiment with Baker indicate that the dog is stupid? _____

No, there is no evidence here of the dog's intelligence.

53

Did Baker lose his ability to act in his own self interest due to the pain of the electric shock? _____

No. Alpha also suffered the electric shock.

54
If Baker would have accidentally found a way to escape the cage after twenty-four minutes, do you think he would try for just one more minute the next time he was placed in the box? _____

No. He is likely to try even longer a second time because of his first escape.

NEW IDEA

55
When Baker quit trying to escape, he had given up hope of effecting his escape. Did Baker have any control over his environment? _____

No. The experimenters controlled Baker's environment.

56
When Alpha found that bumping the hinged side of his cage would allow him to escape, did the dog have control over his environment? _____

Yes

NEW IDEA

57
The **major** cause of anxiety is **loss of control** over the environment. An individual who believes, correctly or incorrectly, that he/she exercises control over the environment will not be anxious. For most people there are portions of the environment they feel they control. Will individuals have anxiety over those portions of the environment they feel they control? _____

No

58
Most people have portions of their environment over which they feel, correctly or incorrectly, they have no control. Will these individuals be anxious about those portions they do not control?

_____

Yes

59

At the end of the experiments, Alpha had
control over the box environment. If
Alpha could tell you how he felt about
himself and the box what might he say?

_____

"That box is easy for me."

"I don't mind that box."

"I like its challenge."

60

At the end of the experiments, Baker had
no control over the box environment. If
Baker could tell you how he felt about
himself and the box, what might he say?

_____

"I hate that box."

"I never want to be put in
    that box again."

CHAPTER 2   QUESTIONS

1. A child was having trouble with math in elementary school. The mother told the child, "I always had trouble with math, too. Maybe it runs in the family." Why is that statement false and misleading?

2. Some individuals are more prone to anxiety than others. Why?

3. Can pain be present during a successful learning experience?

4. An individual in an anxiety creating situation is urged to be active rather than passive. What is the relationship between being active and maintaining some control over the environment?

5. Does active behavior also have to be purposeful behavior?

6. What is the relationship between intelligence and anxiety?

7. If a full year elapsed and the dogs were brought back to the box, what do you think would be their initial reactions?

CHAPTER 2   GROUP LEARNING ACTIVITIES

1. Suppose that Alpha and Baker met in a kennel two days after the experiments. Write a dialogue that might occur between them.

2. Divide a group into "Alpha dogs" and "Baker dogs." Engage in an impromptu, creative drama that might occur shortly after the experiments were completed.

Chapter 3

Re-Constructing a Glass Box to Find a Hinged Side

In Chapter 2 the story of Alpha and Baker was used to illustrate the type of situation that can teach a dog, or a human being, to be anxious. An important understanding you should have acquired from the dog experiments is that anxiety is learned. Math anxiety isn't inherited. Math anxiety is not caused by a lack of intelligence. And anxiety is not a natural reaction to mathematics. Math anxiety is learned and part of the strategy for overcoming math anxiety is recognizing that fact. Frequently, the best way to understand your anxiety is to reconstruct the circumstances under which you learned it.

One important factor in the reconstruction of an individual's anxiety producing situation is the idea of a **"significant person."** This significant person may be a "villain" or a "loved one." The following true stories are intended to illustrate this idea of a "significant person."

MAXINE'S TRUE STORY

Maxine **learned** to be anxious in the third grade. Her teacher was teaching multiplication facts to the class and wanted the students to respond quickly and accurately. As in any speed contest, whether the race involves cars, feet, or mouths, some students respond more quickly than others. Maxine was among the slower students to respond and for her "crime" was forced to sit under her desk while the rest of the class ridiculed her.

Maxine is now over fifty years old. She has proved her competence in many ways. She has an exceptionally good memory. In fact, she remembers sitting under her desk as if it happened yesterday.

Maxine has one terrible problem. She suffers severe anxiety in any math learning situation. She never has become comfortable with the simplest of multiplication problems.

It is difficult to assess the motives of Maxine's third grade teacher. Maybe the teacher was just plain mean - a villain by any standards. More likely, the teacher believed that making an example of Maxine would make her and the rest of the class work hard and learn their multiplication tables. We don't know the effect on other members of the class, but Maxine seems scarred for life. Apparently, sitting under her desk with classmates jeering was a glass box without a hinged side.

ΠΠΠΠΠΠΠ   ΠΠΠΠΠΠΠ   ΠΠΠΠΠΠΠ   ΠΠΠΠΠΠΠ   ΠΠΠΠΠΠΠ

CINDY'S TRUE STORY

Cindy had a very different learning experience. She was excellent in grade school arithmetic and her first year of Algebra, but she had some normal confusion in proving geometry theorems. She asked her father for help and they sat down for a study session.

"Nobody ever expected me to do well in mathematics before."

Cindy loved and idolized her father. When he became quickly exasperated at her questions she was devastated.

The nightly lessons continued, but Cindy's math performance plummeted. As the school term progressed, Cindy's father frequently expressed the opinion that ". . . if we can just get through this geometry then we can turn our academic endeavors elsewhere." Cindy got the message and took no more mathematics.

Today, twenty years later, Cindy regrets that decision. It has eliminated many of her options for further education and/or career. She remains uncomfortable with mathematics.

ɪɪɪɪɪɪɪɪɪɪɪɪ   ɪɪɪɪɪɪɪɪɪɪ   ɪɪɪɪɪɪɪɪɪ   ɪɪɪɪɪɪɪɪɪɪ   ɪɪɪɪɪɪɪɪ

SIGNIFICANT PERSONS FOR MAXINE AND CINDY

For Maxine, the significant person in learning her anxiety was the teacher — a person Maxine never loved or admired. The teacher was and is a "villain" in Maxine's perspective.

For Cindy, the significant person was her father — a person she loved and adored. Cindy continues to love and adore her father, but realizes that her present anxiety is traced directly to the interactions between them many years ago.

In both cases, the significant person seemed to exercise complete control over the environment. If the glass box environment had a hinged side, neither Maxine nor Cindy was able to find it.

LOOKING AT YOUR OWN HISTORY

The stories of Maxine and Cindy are not unusual. Every person who suffers from math anxiety has learned that anxiety reaction because of something that occurred in her/his past.

Some people have no clear memory of that traumatic moment when math became scary and frightening. Perhaps such individuals never experienced a dramatic incident and came to their beliefs through continuous, subtle messages that told them they were inferior in mathematics. When anxiety is learned in an undramatic, unconscious way, there is a high likelihood that the "anxiety classroom" was in the home and the "teacher" was one or more of the family members.

Some people who have no clear idea of how they acquired their math anxiety may have experienced a single, traumatic incident, but the mind has gently and kindly pushed it away from conscious memory. It is not suggested here that uncovering such memory without professional assistance is desirable or necessary.

It is desirable, as a first step toward reducing your anxiety, to put into writing some of your history. If you can easily remember one event that seemed to create your anxiety, then a word description of it will be helpful. If you can not remember a single, traumatic event then probe no deeper; instead, try to recall the first time you realized that you were uncomfortable in a mathematics situation. Whatever your circumstances your history provides important clues to your present and future. The following questions will provide an outline for you to remember, think, and write about your "anxiety history."

CHAPTER 3   QUESTIONS

1. Where did you acquire your anxiety? Include here a complete, physical description of the place, its normal sounds, sights, and smells.

2. No artistic quality is needed for this item because no one will see your work (unless you choose to share it). Draw a picture of the place where you acquired your anxiety. Use colors and be creative; in other words, don't try to draw the scene as it actually was, but let feelings of the place shape your picture.

3. What were you like when you first realized that something seemed to be wrong between you and mathematics? Include here a complete physical description of yourself (age, height, weight, etc.), the clothes you wore, and the feelings you had.

4. Draw a picture of yourself as you were when you acquired your anxiety. Again, be creative. Use colors. Maybe you can draw yourself into the picture of the place drawn in Question 2.

5. Who were the significant people involved? Include here a description of your own feelings toward the other person(s), any relationships that affect the feelings, and any sights, sounds, or smells you associate with the person(s).

6. Draw a picture of the significant people. Be creative. Use colors. Put them in the place of Question 2 if possible.

7. What happened? Include here any dialogue, facial expressions, or other communications that seem relevant.

8. Do you think your experience was comparable to the glass box? If so, do you think your experience had a hidden escape mechanism like Alpha's box or no escape like Baker's box?

CHAPTER 3   GROUP LEARNING ACTIVITIES

1. Share individual histories of how math anxiety was acquired.

2. Discuss the question:

   Is it helpful for a math-anxious person to find
   other people that have similar feelings?

3. Discuss the question:

   How is an individual's history going to help in
   reducing the math anxiety?

Chapter 4

Gaining Control in School Situations

A NEW LIFE FOR BAKER? ? ? ? ?

The story of Baker in Chapter 2 is incomplete because it seems to leave the dog ruined for life by the experiment. That need not be the way the story ends because there are other experiments for Baker which may offset or greatly reduce the damage done by the glass box experiment. Just as the glass box experiment was designed to create anxiety, other ways of treating Baker can be designed to overcome, at least partially, the anxiety.

ꟷꟷꟷ   ꟷꟷꟷ   ꟷꟷꟷ   ꟷꟷꟷ   ꟷꟷꟷ

GAINING CONTROL OVER YOUR ENVIRONMENT

The most crucial factor in creating anxiety or reducing its effects is control. Baker's anxiety was created in a situation where he exercised no control over his situation. Baker tried urinating, barking, jumping, whining, running in circles, and bumping the sides of the box; none of these behaviors had any effect (good or bad) on his situation. Baker had no control and anxiety is created when the person (or animal) feels there is no way to control the significant factors in his/her environment.

41

The dog Alpha experienced pain in the glass box, but also found a way of controlling the environment. Alpha knew how to escape the glass box and this ability gave him control over the situation.

You are probably reading this book because you have learned to control important segments of your life and are willing to test your ability to control mathematics. The preceding chapters of this book will help you gain control over some of the self-defeating myths like:

—"Some people are naturally smart, and others naturally dumb, in mathematics."

—"My mother (father, sister, brother, etc.) was never good at mathematics either."

Gaining control by understanding that you have no natural, biological, intellectual deficiency is an important step, but it offers no direct help for the person who enrolls in a math class and finds her/himself confronting situations and individuals (teachers and students) that are similar to the environment and significant people where anxiety was created in the past. Such a person needs new control mechanisms and understandings.

ΙΙΙΙΙΙΙΙΙΙΙ   ΙΙΙΙΙΙΙΙΙΙ   ΙΙΙΙΙΙΙΙΙΙ   ΙΙΙΙΙΙΙΙΙΙ   ΙΙΙΙΙΙΙΙΙΙ

LEARNING TO LEARN

This chapter is a short course on learning and instruction. If you understand how an individual learns and how good instruction facilitates that process then you can apply that knowledge to gain control over school situations. This is important for two reasons:

1) Anxiety sufferers frequently are hindered in their learning because their behavior is inappropriate for the tasks, and

2) Anxiety sufferers generally blame themselves for their failures even in those situations where poor instruction is the major culprit.

The objective of this chapter is to give you control over your learning. You will know when you are using an effective process for learning. You will know when you are receiving good instruction that will facilitate your learning. You will exercise control over your own learning environment.

Again, the information of this chapter is given by the question-answer format of Chapter 2. As the lesson unfolds, the rationale for this format should become obvious. As before, read each numbered paragraph carefully, write an appropriate response for its question, and then look immediately at the answer to see if you understood correctly.

1
Anxiety is created when an individual feels she/he has no control over the important factors in her/his environment. Is physical pain the cause of anxiety? _____   No

2
Anxiety is always future-oriented because the individual perceives a situation as threatening. The individual's past history has taught her/him to feel there is no control. Past history leads the individual to the belief that she/he has no control over the environment, but anxiety is
_____ -oriented.   future

3
Fear of what will happen is a symptom of anxiety. If an individual had an anxiety producing experience in a math classroom, then entering another math classroom is likely to be a _____ (secure, threatening) situation.   threatening

NEW IDEA

4

One way to reduce classroom-related anxiety is to gain control over the learning process. In most formal classroom situations, the teacher is seen as controlling the learning process. For the anxiety sufferer, who should control the learning process?

_____

the anxiety sufferer

5

The individual who does not understand the learning process is dependent upon an instructor in a formal, classroom situation. The individual who understands the learning process _____ (can, cannot) exercise control in a formal, classroom learning situation.

Can (sometimes this control is limited to understanding why a concept is not being learned, but this understanding shifts the control to the learner).

6

A student exercises control over his/her learning environment by understanding the learning process. Is control an essential factor in reducing anxiety? _____

Yes

7

The next series of question-answer items (frames) is designed to give you an understanding of the learning process. With this understanding you can exercise _____ over learning situations.

control

8

There are three clearly identifiable steps involved in the learning process. How many steps are involved in learning? _____

three

NEW IDEA

9

The first step in the learning process is
awareness. The learner must be **aware** of something
new or different in his/her environment. Can
learning occur when a person perceives only known,
familiar elements in her/his environment? _____   No

10

Awareness is the first step in the learning
process and no learning can occur without it.
One aspect of awareness is the necessity for an
adequate background. To be aware of a new concept
in a calculus course, the student needs _____
(more, less) than an ability to do arithmetic.   more

NEW IDEA

11

An adequate background is a necessity for being
aware. Another necessity for awareness is the
ability to **consciously focus** on the new idea,
skill, etc. that is to be learned. A frequent
cause of failure to learn is the inability of
the student to consciously _____ on a new idea.   focus

12

In classrooms, the student often pays
attention to information that is not
directly related to the lesson. If the
teacher is trying to teach the multiplication symbol (x) and uses the
example

$$5 \times 3 = ?,$$

what happens to the student who concen-
trates on the "5" and the "3?"

_____

The reader who thought the correct answer to this frame is
15 proves the frame's lesson
which is: The student is not
aware of what is to be focused
upon (learned).

13

Learning can only occur after a person is _____
of some idea, skill, etc. that is new or different
in his/her environment.

aware

14

Awareness is the first step in the
learning process and always requires:

1) An adequate background, and

2) _____

An ability to consciously
focus on the new idea, skill,
etc. that is to be learned.

15

Suppose: You awake at night and hear a
strange noise in the house. You begin
listening intently to the sound. Have
you completed the first step in the
learning process? _____

Yes. You are aware of something new in your environment
and are consciously focusing
upon it.

16

Suppose: After hearing the strange sound
that awakes you, you go back to sleep.
Have you learned anything? _____

No. Awareness is only the
first step in the learning
process.

NEW IDEA

17

Awareness is the first step in the learning
process. The second step requires some type of
active response from the would-be learner. If
you hear a strange noise and get out of bed to
investigate it, are you **actively responding** to an
awareness? _____

Yes

18
The second step in the learning process is **action-response.** If you hear a strange noise and lie in bed figuring out what the noise is like, are you actively responding to an awareness? _____

Yes. Action-response is not limited to physical movement. The mind can actively respond to an awareness.

19
In a classroom, a teacher was showing how to do a new type of problem and asked the class to work a similar problem. One student was afraid of doing it wrong and drew a picture instead. Was the student's behavior an appropriate **action-response** for the lesson? _____

No. Good learners often get wrong answers. The action-response step of the learning process demands that the learner take the risk of trying ideas and skills that are not yet completely understood.

20
An appropriate action-response is the second step to the learning process. What is the first step?

_____

Awareness

21
A child sees a bug and describes it with words. Have both the awareness and action-response steps of the learning process been completed? _____

Yes

22
A child sees a bug and then draws a picture of the bug. Have both the awareness and action-response steps of the learning process been completed?

_____

Yes

23
A teacher describes how to do a problem and then shows three examples. Have both awareness and action-response been completed? _____

Probably not. Action-response must be taken by the learner.

24
A teacher describes how to do a problem and then the student works an example. Have both awareness and action-response been completed? _____

Probably so. Yes.

25
There are three steps to the learning process. When the teacher shows how to do a problem and the student works an example, are all three steps completed? _____

No. Only awareness and action-response.

NEW IDEA

26
The third step in the learning process is **feedback.** The learner needs to receive some information from his/her action-response. This information on the accuracy or appropriateness of action-response is _____ .

feedback

27
The third step is feedback and the first two steps are _____ and _____ .

awareness, action-response

28
A strange noise wakes you and you go to investigate it. You find the broom leaning against the refrigerator and vibrating when the motor runs. Have you completed the three steps in the learning process? _____

Yes

29
After a math class lecture, you go home and work thirty problems. The next day you go over the problems in class. Have you completed the three steps in the learning process? _____

Yes

30
The three steps in the learning process are:

    1)   awareness,

    2)   action-response, and

    3)   feedback.

Can learning occur if only two of the steps are completed? _____

No

31
The three steps in the learning process are _____, _____, and _____.

awareness (focusing),
action-response,
feedback

NEW IDEA

32
Most important learning tasks are composed of a series of sub-tasks or intermediate parts. In the long division of

$$43 \overline{)17,634}$$

will there be a series of smaller tasks that need to be accomplished? _____

Yes. This one problem requires a number of sub-skills. Three of those sub-skills are: estimation, multiplication, and subtraction.

33
Almost every learning task requires some sub-tasks and each of these sub-tasks must be mastered before the major task can be done correctly.

$$\begin{array}{r} 5,673 \\ \times\ 16 \\ \hline \end{array}$$

Can the multiplication shown above be completed in a single step? _____

No

34

In trying to learn a difficult concept or procedure will there be only one idea or skill to be mastered? _____

No

NEW IDEA

35

To learn a difficult concept or procedure, first identify as many separate sub-tasks as possible. This accomplishes two desirable objectives.

    1)   It improves the likelihood that the awareness step (focusing) has been successfully completed.

    2)   Each sub-task is simpler and, therefore, easier to learn.

If a difficult learning task is encountered, break it into ____ - _____ .

sub-tasks

36

If a learning task is easy, then learn it all at once. But if a learning task is difficult, break it into separate _____ _____ .

sub-tasks

NEW IDEA

37

Break a difficult concept or skill into smaller lessons or sub-tasks. Eventually, however, it is necessary to integrate those sub-tasks and learn the original concept or skill as a single unit. If you learn separate sub-tasks, must you also learn the relationships between them? _____

Yes. Difficult concepts or skills may be separated, for the purposes of learning, into smaller and simpler sub-tasks, but those lessons need to be integrated and their relationships understood.

38
The learning of any subject, like algebra, is
accomplished by learning separate skills and then
finding relationships between those skills.
A subject like algebra is a complex learning task
that is best learned by breaking it down into
smaller _____ _____.   sub-tasks

39
Learning a lesson, like long division, is best
accomplished by learning its separate sub-tasks
and then integrating them into a single unit.
When a lesson seems easy, it _____ (should,
shouldn't) be broken down into sub-tasks.   shouldn't

40
When a lesson seems difficult, it _____
(should, shouldn't) be broken down into sub-tasks.   should

NEW IDEA

41
Research on the learning process indicates
that a person can only handle about six
separate lessons (sub-tasks) in a single
lesson. If a lesson requires ten separate tasks, should the student attempt to
learn them all in one study session?

_____   No. Learn no more than six
                                                           in one session. Then integrate
                                                           them into your other knowledge
                                                           before learning the other four.

42
Research indicates that a student should try no
more than ___ separate lessons before integrating
them into a a single, complex concept or skill.   6

43
Each new learning task requires three steps.
They are: awareness, _____,   action-response,
and _____.   feedback

51

NEW IDEA

44

The learning process is most effective when each step begins immediately following its predecessor. If action-response occurs on Monday and feedback is given on Wednesday, is this desirable? _____

No. Feedback needs to follow action-response immediately.

45

A student attends a math lecture on Thursday morning and does the homework that evening. How might the student alter her/his schedule to improve the use of the learning process?

_____

Do the homework immediately after math class so that action-response immediately follows awareness.

46

After finishing his/her math homework problems on Thursday, the student waits until class the following Tuesday to see if the problems were done correctly. What needs to be done to improve the use of the learning process?

_____

Answers need to be checked immediately after the problems are completed. Feedback needs to follow action-response immediately.

47

Teachers sometimes design their instruction so that there is a lengthy time separation between awareness and action-response. Are such teachers making learning as easy as possible for their students? _____

No

48

Math textbooks frequently include answers for some, but not all, of their problems. Are such textbooks designed to make learning as easy as possible? _____

Probably not. Unless feedback can be obtained from another source, the learning process cannot function effectively.

49

The three steps in the learning process are:
1) awareness,
2) action-response, and
3) feedback.

The process works best when each awareness is followed immediately by its _____.

action-response

50

If a math problem requires four steps in its solution then there are four separate sub-tasks. Some teachers will show how to do the whole problem and then ask the students to work a similar problem. What would be a better approach in teaching this type of problem? _____

Teach it as four separate sub-tasks, allowing for awareness, action-response, and feedback on each, and then integrate those four sub-tasks into a single procedure.

51

In college math classes the professor often lectures a full hour without interruption for questions or student work. Are such lectures designed with an understanding of the learning process? _____

No. A lecture is designed only to meet the awareness phase of learning. Furthermore, after the lecture covers more than six sub-lessons, all further material is likely to miss even the awareness phase of the process.

52

It is possible to understand the learning process and still learn something that was unintended or false. Sometimes, a student becomes aware of an irrelevant attribute. Might this misplaced awareness create a problem in what is learned? _____

Yes. The student's action-response may be inappropriate and/or the feedback mis-interpreted.

53

A student who is aware of an irrelevant attribute may learn something that was unintended. Might a student also fail to learn by choosing an inappropriate action-response? _____

Yes. The feedback will quite likely be misleading.

54

Sometimes a student will correctly apply both awareness and action-response, but the feedback is misleading. Could this occur when the math text has the wrong answer to one of its problems?

_____

Yes

55

A failure to learn can occur whenever something distorts any step in the learning process. If a student doesn't understand directions might that student fail to learn the desired lesson? _____

Yes. Awareness is distorted.

56

In any formal learning situation, someone should be responsible for maintaining a good learning process. In classroom situations, the teacher should be responsible, but some students naturally accept this responsibility also. Do you believe that these students are successful in math courses? _____

Yes. Students who accept responsibility for maintaining a good learning process are often considered to have high intelligence. Actually their intelligence isn't the difference; it's their use of the learning process.

NEW IDEA

57

Research on excellence in instruction shows that four factors are always followed by good teachers. The first factor is that an excellent teacher gives clear directions or cues. How is that factor related to the learning process? _____

Clear directions or cues will make students aware (focus).

NEW IDEA

58

Excellence in instruction requires four
factors. The second factor is that the
good teacher always designs activities
that will actively involve the students.
How is this factor related to the learning process? _____

The good teacher has
planned an appropriate action-
response by the students.

59

Research has shown that there are four
factors involved in excellent instruction.
The first two factors are directly related
to _____ and _____ which
are the first two steps in the learning
process.

awareness (focusing),
action-response

NEW IDEA

60

The third factor in excellent instruction
is feedback. This is _____ (identical
to, different from) the third step in the
learning process.

identical to

61

Excellent instruction must make allowance
for the possibility that learning will not
occur (there are many interferences in the
learning process). The fourth factor in
excellent instruction requires corrective
strategies in those instances where the
instruction failed to meet its objective.
Does the excellent teacher accept responsibility for overcoming those instances
when the student fails to learn? _____

Yes. Excellence in
instruction accepts the fact
that failure to learn will
sometimes occur and plans for
new instruction in such
situations.

CHAPTER 4   QUESTIONS

1. In a college lecture class, what can a student do during the lecture so that the situation will better fit the learning process?

2. If a teacher gives homework, what should the student do to make the situation better fit the learning process?

3. If the answers to homework problems are not readily available, what can the student do so that the situation will better fit the learning process?

4. If a teacher completely ignores the learning process, who is responsible for student failure?

5. "Street smart" is a term used to describe the wise behavior of many youngsters in their neighborhoods. What might the term "school smart" mean?

6. Select a mathematics teacher from your past or present. Evaluate the teacher on how well he/she seems to understand the learning process.

CHAPTER 4   GROUP LEARNING ACTIVITIES

1. Have the members of the group list all the possible sub-tasks involved in:
   a. The long division of whole numbers.
   b. The long division of decimal numerals.
   c. The addition of two fractions.
   d. The division of two mixed numbers.

2. Have each member of the group choose a simple learning task such as sewing a hem, changing an automobile tire, or making cocoa. Each member designs a learning process and teaches the task to the group with discussion afterwards on the learning process and teaching.

Chapter 5

Four Necessary Mathematics Learning Strategies

Three steps in the learning process and four qualities of excellent instruction were presented in Chapter 4. They prepare you to diagnose correctly the point at which you experience difficulty in your study of mathematics. For example, if you find yourself unable to focus your attention on a particular lesson then you should know that the **awareness** step of the learning process is not working for you. That knowledge, however, needs to be supplemented by another set of strategies and mathematics study skills for you to go beyond diagnosis and find a way to focus upon the new material.

This chapter describes four areas of needed mathematics study strategies. These strategies are adapted from the work of Dr. Claire E. Weinstein, professor of Educational Psychology at the University of Texas. Weinstein cautions that these four learning strategies must not be practiced separately. She compares them to the skills of driving a car. It is possible to list some driving skills separately (shifting, steering, braking, etc.), but a good driver must have all of the skills. Similarly, all of the mathematics learning strategies are absolute necessities.

The lack of any one will severely limit any progress at really learning mathematics.

ㅠㅠㅠㅠㅠ   ㅠㅠㅠㅠㅠ   ㅠㅠㅠㅠㅠ   ㅠㅠㅠㅠㅠ   ㅠㅠㅠㅠㅠ

FOUR NEEDED MATHEMATICS STUDY SKILLS

We'll list the four areas of mathematics study strategies and then further explain them.

1. Information acquisition

2. Maintaining a supportive environment

3. Using active learning strategies

4. Monitoring comprehension

ㅠㅠㅠㅠㅠ   ㅠㅠㅠㅠㅠ   ㅠㅠㅠㅠㅠ   ㅠㅠㅠㅠㅠ   ㅠㅠㅠㅠㅠ

1

The learning of mathematics requires the handling of two types of information. Some of this information needs to be memorized. Some must **not** be memorized. If possible, should you memorize all the information in a mathematics course?

No

NEW IDEA

2

In mathematics, the meaning of every symbol must be memorized. Symbols such as %, +, and $\sqrt{}$ have no meaning inherent within them. Is it possible to figure out the meaning of the symbol "#" in the following sentence?

$$5 \ \# \ 7 = 31$$

No. Most symbols have no meaning that can be "figured out." The meaning of a symbol in mathematics needs to be memorized.

3

Most symbols have no meaning that can be easily discovered. The meaning of a symbol in mathematics needs to be _____ (figured out, memorized.)

memorized

4

The meaning of a symbol is one type of information that needs to be memorized. Another type of information that must be memorized is the **definitions** of technical words. There is nothing about the word "parallel" which describes its meaning. Can you discover the meaning of the word "trapezoid" by studying the word carefully?
_____

No

5

Words usually have no relationship between the way they look and their intended meaning. Definitions in mathematics must be _____.

memorized

6

In mathematics, always memorize the meaning of any symbol or definition. Should everything in mathematics be memorized? _____

No

7

Symbols and definitions comprise only a small part of mathematics and must be memorized. Should everything in mathematics be memorized? _____

No

NEW IDEA

8

Any **process** in mathematics must be understood before it has been learned. Examples of **processes** are:

- adding columns of numbers

- solving algebra equations

Is it sufficient to memorize a process in trying to learn mathematics? _____

No

9
Processes in mathematics need to be understood.
This understanding is possible because processes
always have a logic inherent in them that makes
it possible to "see" what is happening. Is it
possible to figure out a process? _____

Yes

10
When a process is to be learned and it seems
impossible to do that, the difficulty is:

  — an inadequate background,

  — an inability to focus, or

  — anxiety.

Does every math process contain a logic that can
be understood? _____

Yes

11
Students who have trouble with mathematics
frequently believe that mathematics is "a
bunch of rules to be memorized." Is that
the way you have described mathematics?

_____

If your answer is "Yes" then
we have just found one of your
serious misunderstandings and,
perhaps, your major difficulty
in learning mathematics.

12
Rules in mathematics were never intended to be
mysterious secrets that are blindly followed.
Rules in mathematics should be _____
(memorized, understood).

understood

13
Rules and processes in mathematics need to be
understood to properly learn the subject. The
meanings of symbols and definitions must be
_____ (memorized, understood).

memorized

14
Rules and processes in mathematics must be
_____ (memorized, understood).

understood

15
Recall from our earlier study of learning
theory that a student needs to concentrate
on a limited number of new ideas (about
six) until they are completely learned.
Does this mean that a teacher must be
careful of the number of new symbols and
definitions that are introduced in a
single presentation? _____

Yes. No single, instructional
presentation should include
more than six new symbols
and/or definitions.

16
Processes or rules are to be understood, but frequently processes are memorized. Does it make
sense to memorize a process that has more than
six steps? _____

No

17
In learning a process, always look for the logic
that connects two steps. Is there always a logic
to a process or procedure in mathematics? _____

Yes

18
Try not to memorize the order of the steps in a
mathematics process, but look at each step and ask
these types of questions:

Why does this step occur here?

What does this step accomplish?

Where did this come from?

In learning a process, is there always a logic
between two steps? _____

Yes

19
It is not possible to "discover" the meaning of
a symbol or definition. Is it possible to "figure
out" the reasons behind a process? _____

Yes

NEW IDEA

20

Every process or rule in mathematics has a logic to it and careful study will always uncover the reasons for it. If a process has ten steps in it, can that process be learned in a single class time? _____

Yes. The limit of six applies to separate, disconnected pieces of learning, but a process, when learned correctly, becomes one, single lesson.

NEW IDEA

21

A second needed study skill for mathematics requires that the learner must work in a supportive environment. Is a darkened room a supportive environment for a reading assignment? _____

No

22

The student must study (work) in an environment that supports rather than distracts or negates progress. Loud music is not a characteristic of a supportive environment. Might soft music be part of a supportive environment? _____

Yes

23

Soft music may not be distracting in a quality study environment. Might soft music cover other noises in the study environment?

_____

Yes

24

Background music with a relatively low number of beats per minute (about 50) can lower your heart beat. Music with a relatively high number of beats per minute can be expected to _____ (raise, lower) your heart beat.

raise

NEW IDEA

25
Soft background noise with a low number of beats per minute is conducive to a good environment for study. The best noise for a good study environment will cover other sounds, lower the student's heart beat, and allow the student to concentrate on the _____ (lesson, sound).

lesson

26
There are two types of good study environments to be considered. The external environment includes items such as lighting and work space. Would the temperature be part of this external environment?

_____

Yes

NEW IDEA

27
Besides the external study environment, there is an internal environment. Within each student there are conditions that will be conducive to good study. If a student is worried about a financial problem, might it be impossible to focus on a mathematics assignment? _____

Yes

28
Among the most distracting characteristics of an internal environment is a feeling of anxiety. Is anxiety likely to decrease the quality of a student's internal environment? _____

Yes

29
Teachers and parents have always emphasized the necessity for studying in a high quality external environment. The internal environment is as important as the external environment. Is the external environment more important than the internal environment? _____

No

30
Before a student begins to study, both the external and _____ environments need to be carefully prepared.

internal

NEW IDEA

31
Is it reasonable to expect a student to
overcome poor lighting or distracting
noise by "studying harder?" _____

No. The external environment
needs to be improved **before**
study begins.

NEW IDEA

32
Is it reasonable to expect a student to
overcome fear, anxiety, or worry by
"studying harder?" _____

No. The internal environment
needs to be improved **before**
study begins.

33
A third type of needed study skill for mathematics
is the use of strategies that will make the student mentally and physically active. Learning
_____ (is, isn't) improved by experiencing sights,
tastes, feelings, smells, and sounds.

is

34
Some individuals are apparently able to remain
completely passive physically while being very
active mentally. Might a person who appears to
be asleep actually be thinking deeply? _____

Yes

35
Most students can attain a good quality of mental
activity by choosing an appropriate physical
activity. Might note-taking be a well-chosen
physical activity to keep a student's mind active?
_____

Yes

36
Some people are able to think well while appearing
to be almost asleep, but such models are completely
unrealistic for most students. The more _____
(active, passive) a student is physically, the more
likely that the student will be active mentally.

active

37
An active mind is necessary for learning. Usually the mind will be more active if the _____ is also active.

body

38
The more active mentally a person can be the more likely that individual is learning. Your five senses (taste, touch, sight, smell, sound) may increase your _____ activity.

mental

39
The third needed mathematics study skill is the use of strategies that will make the student mentally active. Learning depends upon a mind that is _____.

active

40
Some individuals are able to maintain a completely passive physical appearance when their mind is very active. Are you able to sit quietly and think about mathematics? _____

If you can answer "Yes" you are a rarity.

41
Might it help your learning to make a pencil drawing? _____

Yes. That may make your mind active.

42
To make your mind active, use:

> **Analogies** — think of other situations that work in similar ways.

If you are trying to add two fractions with different denominators, might it help to think of how to add fractions with the same denominator?

_____

Yes

43

To make your mind active, use:

>**Examples** — number (other) applications
>that show specific cases in which some
>generality works.

Would it be helpful for understanding the formula
for the area of a square to show how it works with
a square with 4 inch sides? _____   Yes

44

To make your mind active, try:

>**Alternative Processes** — Find other steps
>that would work in solving a problem.

If you look for another way to correctly do a
problem, will that help keep your mind active?

_____   Yes

45

To make your mind active, try:

>**Contradictions** — Situations where this
>learning does not seem to apply.

To decide whether every number is either positive
or negative, might you learn something by finding
a number that is neither?   Yes.   Zero (0) is
                                                             neither.

46

Keep your mind active by asking yourself good
questions. Try to ask "what," "why," "where,"
questions and avoid "how" questions. A "why"
question suggests an _____ (understanding,
memorization) is sought.   understanding

47

"How" questions suggest there are steps to be
memorized, but "what," "where," "why" questions
suggest a need to _____ .   understand

48

"How" questions suggest there are steps to be
memorized, but "why" questions suggest a need to
understand. When learning a process don't ask
_____ (how, why) questions.   how.   Processes
                                                             are to be understood
                                                             rather than
                                                             memorized.

49

In trying to become an active learner, always work
toward imagery. If the student can close his/her
eyes, mouth, ears, etc. and imagine all those
senses at work on the problem then it is likely
that the mind is active. If you start dreaming
of mathematics problems is it probably a sign
that you are becoming a good math student? _____   Yes

NEW IDEA

50

The three mathematics study skills already dis-
cussed in this chapter are:

    1   Memorize the meaning of mathematics
       words and symbols, but understand
       all processes.
    2   Prepare both your external and
       internal environments to provide
       optimum study conditions.
    3   Be an active learner.

The fourth mathematics study skill you must acquire is an ability to effectively use the feedback you receive. Is feedback an important factor
in learning? _____   Yes

51

The student must become adept at:

    1   Demanding feedback on the quality of
       her/his responses, and
    2   Using the feedback to consolidate
       and/or alter his/her understanding
       of the material.

If a student has no idea of the correctness of an
answer to a mathematics problem, has quality
learning occurred? _____   No

52

A student finds that he/she has done a problem
incorrectly. Which of the following actions
should be taken?

    a) Go to the next problem and try
        to get it correct.
    b) Find the mistake and alter the
        process to avoid making that mistake
        again.

_____   b

53
A student does a problem and finds the answer is
correct.   Which of the following actions should
be taken?

      a)   Go to the next problem.
      b)   Go over the problem and review the
          correct process.

_____   b

54
When doing mathematics problems, both correct and
incorrect answers have appropriate responses.
When an answer is incorrect the student needs to
respond by:

      a)   Finding the mistake and understand-
          ing how to avoid it in the future.
      b)   Feeling stupid.

_____   a

55
When a student does a problem and gets the correct
answer, she/he should:

      a)   Feel good about it.
      b)   Review the process and recognize the
          logic for it.
      c)   Both of the above.

_____   c

56
Frequently mathematics students do not seek
(demand) an answer for every problem completed.
Should a student always receive some indication
of the correctness of her/his answer? _____   Yes

57
Using feedback wisely is called **good comprehension
monitoring.** Good comprehension monitoring
requires demanding answers to _____ (some, all)
of the problems that are worked.   all

58
Good comprehension monitoring requires the learner
who gets a correct answer to accept it with:

      a)   A matter-of-fact attitude.
      b)   A conscious feeling of pride in the
          accomplishment.

_____   b

59
When an incorrect answer is obtained, good comprehension monitoring requires:

    a) A willingness to push ahead with the hope that the difficulty will be understood later.

    b) A reluctance to proceed further until the mis-learning that created it is understood.

                    b

_____

60
Good comprehension monitoring requires the learner to accept primary responsibility for evaluating her/his achievements. A student should:

    a) Accept the teacher's grades as accurate.

    b) Use the teacher's grades to help in making an evaluation of the quality of achievement.

                    b

_____

61
Good comprehension monitoring demands that the student _____ (is, isn't) responsible for her/his academic performance.

                    is

CHAPTER 5   QUESTIONS

1. Name the four needed mathematics study skills.

2. Why can the four needed mathematics study skills be considered similar to the skills necessary for driving an automobile?

3. One method of creating an active strategy for studying is the use of analogies. What is meant by "analogies?"

4. What type of mathematics should be memorized?

5. What type of mathematics must be understood?

6. Why should a student avoid asking questions such as:

    How is this problem worked?

CHAPTER 5   GROUP LEARNING ACTIVITIES

1. Observe an individual or a group studying mathematics. Write explicit descriptions of the use (or non-use) of the four needed learning strategies.

2. Try conducting a group meeting with loud, fast music in the background.

3. Try conducting a group meeting with soft, slow music in the background.

4. Select a mathematics process that has been memorized. Find the logic inherent in that process.

Chapter 6

Calming Math Fears

Fear plays a critical role in the survival of most animals because it marshals the animal's strength and makes it more powerful in a fight or stronger as it flees. At some time in the history of human beings, fear also played a critical role in survival, but those times have past and human beings are no longer well-served by their body reactions to fear. This is because few of the fears that are experienced are related to physical survival.

Most fears in an industrial society are fears of psychological well-being. An employee experiences fear when the boss terminates employment. A man experiences fear when the woman he loves rejects him. A student experiences fear when a teacher demands the answer to a difficult question. In all of these cases the fear evokes physical responses which may have been appropriate for a caveperson ancestor, but not for today's human being.

What value is it for the person losing his job when his heart beats wildly preparing him for physical exertion? What value is it for the woman losing her mate when adrenaline is released into the bloodstream? What value is it for the student confronted with a difficult question when his lungs and air passages dilate providing extra oxygen and setting off heavy breathing?

Anxiety sufferers almost always experience physical reactions in those situations where their anxiety is aroused. Those physical reactions are inappropriate for the fear and, furthermore, interfere with the individual's ability to respond appropriately. Sweaty palms, a wildly beating heart, severe nausea, and knots in the stomach are common occurrences. Uncontrolled crying, shaking, and vomiting are less common occurrences for anxiety sufferers, but such reactions are not rare.

"How am I ever going to use this stuff?"

Any physical reaction to math anxiety will interfere with one's ability to correctly perform mathematics. This means that the more a person suffers from math anxiety, the greater will be the physical reactions to the fear and the worse will be the individual's math performance. The physical reactions exaggerate the effects of the anxiety. This fact alone would make it worthwhile to develop ways to reduce physical reactions, but there is an additional benefit.

Research has shown that if the physical symptoms of anxiety are reduced then the individual experiences a reduction in the anxiety itself. Therefore, any strategy that will calm an anxiety-sufferer physically will also benefit the individual psychologically and emotionally.

мимими   мимими   мимими   мимими   мимими

BENSON'S RELAXATION RESPONSE

The objective of this chapter is to develop a process for reducing the negative physical reactions to anxiety. The process is called **Benson's Relaxation Response*** because it was formulated by Harvard Professor Herbert Benson, M.D. The Benson Relaxation Response was developed from a study of methods used in a number of cultures. Benson and his colleagues tested these methods and scientifically validated a process that aims to calm and relax any person practicing the technique. Some people report drastic improvements in their relaxation; others report little noticeable change.

But, Benson's research shows that all people practicing his process will have "physiological changes such as decreased oxygen consumption." For the anxiety-sufferers, this means that learning the Benson Relaxation Response promises a reduction in the negative physical reactions to anxiety and resulting reductions in the anxiety itself. Another advantage to the Benson Relaxation Response is that the process, once learned and practiced, can be followed in the classroom when anxiety is aroused; it does not require a special position or behavior that would make it difficult to follow publicly.

The following question-answer format section begins with a review of what has been covered earlier in this book and then teaches the Benson Relaxation Response.

*Benson, Herbert, The Relaxation Response,
William Morrow and Co., Inc., New York, 1975

1
A reduction of math anxiety is accomplished by a coordinated attack against many factors related to it. One attack must debunk some myths about mathematics and intelligence. Is it true that math anxiety is only suffered by individuals who have limited intelligence? _____   No

2
It is a myth that there is a correlation between math anxiety and intelligence. Another myth concerns women and mathematics. Is there anything in the physiological or intellectual makeup of women that makes them less capable in mathematics than men? _____   No

3

Debunking myths is part of the process in reducing
math anxiety. Is it a myth that anxiety is inherited from a parent who never could do math?
_____

Yes, that is a
myth.

4

Psychological experiments with animals have shown
how anxiety is created. Is pain the cause of math
anxiety? _____

No

5

Perceived lack of control is the major cause of
anxiety. If an individual feels threatened and
seemingly has no way to control the threat, is
he/she likely to be anxious? _____

Yes

6

Lack of control is the cause of anxiety. Is
repeated failure the cause of anxiety? _____

No. Continued
failure may
accompany anxiety,
but it is not the
cause of it.

7

One method of gaining control over math anxiety
is to gain an understanding of the teaching-learn-
ing process. Is it a myth that the individual who
understands how to learn gains some control over
school situations? _____

No, this is
not a myth.

8

The learning process consists of three steps:

    1)   awareness,
    2)   action-response, and
    3)   feedback.

Can the student monitor his/her learning
activities and determine whether the three
steps are being followed? _____

Yes

9
The classroom instructor is responsible for the quality of instruction, but the student who understands the learning process can also exercise some control over his/her learning. Is control necessary in reducing anxiety? _____

Yes

10
Four mathematics study skills are needed by any student who hopes to be successful in the subject. Can a student with only three of these skills expect to be successful in an Algebra course? _____

No

11
One study skill for success in mathematics requires a student to treat information in the subject correctly. What type of information must be memorized in a mathematics course?

_____   _____

Definitions;
Meanings of
symbols

12
What type of information must be understood in a mathematics course? _____

Processes, Rules

13
Two types of environments must be carefully prepared by the successful mathematics student. What are those two environments?

_____   _____

External; Internal

14
Name one quality of a high quality external environment. _____

Good lighting;
comfortable temperature; Soft,
slow background
noise; etc.

15
Name one quality of a poor quality internal environment. _____

Anxiety; worry;
other concerns;
etc.

16

In confronting a difficult mathematics question the successful student will be _____ (active, passive).

active

17

Successful mathematics students do which of the following:
- a) Sit quietly and think.
- b) Engage as many senses as possible.
- c) Find examples, alternatives, con-tradictions, etc.

_____

b and c. Few people can sit quietly and think; others need strategies to learn effectively.

18

When a successful mathematics student arrives at an incorrect answer:
- a) He/she tries another problem to see if a correct answer can be found.
- b) She/he finds the mistake and seeks a better understanding of the process.

_____

b

19

When a successful mathematics student arrives at a correct answer:
- a) He/she goes directly to the next problem.
- b) She/he reviews the problem and seeks a deeper understanding of the logic behind the process.

_____

b

20

The four needed study skills for success in mathematics are:
1) _____
2) _____

3) _____

4) _____

Appropriate use of information.
Careful preparation of both the external and internal environments.
Use of active study strategies.
Demanding and using feedback.

21
There is a tendency for a math anxiety-sufferer to blame her/himself for any failure. Is it possible that the blame is misplaced? _____

Yes. Although it is healthy for an individual to accept responsibility for his/her own learning, it is not healthy to assume blame for every failure.

22
The individual that believes that he/she is inept in mathematics will also believe that failure is caused by personal deficiencies. Is it helpful for a person to assume that failure is caused by personal deficiencies? _____

No

23
The individual who believes she/he is capable in mathematics will look to possibilities outside him/herself in the event of failure. Might there be other reasons for failure besides personal deficiencies? _____

Yes

24
Which of the following is a preferable belief for helping a person exercise better control over learning mathematics?

    a) The individual believes he/she is inept in mathematics, or

    b) The individual believes she/he is capable in mathematics.

_____

b

25
A realistic understanding of a situation is often helpful in warding off anxiety. Is it ever helpful to make negative assumptions about yourself or your abilities? _____

No

26
Most math-anxious people experience physical
reactions as part of their anxiety. Sweaty palms
and difficulty in breathing are two physical
symptoms of anxiety. Might a wildly beating
heart be a symptom of anxiety? _____

Yes

27
Uncontrolled crying and vomiting are two severe
physical reactions to anxiety. Does an individual
with severe physical reactions to anxiety have a
reasonable chance to succeed in a math classroom?

_____

No

28
Physical reactions to anxiety interfere with an
individual's ability to do mathematics. If those
physical reactions can be reduced will it improve
the learning abilities of an anxiety sufferer?

_____

Yes

29
Research has shown that processes that reduce the
physical reactions to anxiety also reduce the
anxiety itself. In some illnesses getting rid of
the symptoms simply disguises the sickness. Is
that true with the physical symptoms of anxiety?

_____

No

30
There is a process for reducing physical symptoms
of anxiety. The first time it is tried for some
people it provides noticeable relief. Can the
results be expected to improve with practice?

_____

Yes

31
Some people notice little change in themselves
when they try the **Relaxation Response,** but research shows that valuable physiological changes
actually do occur. Should a person with math
anxiety try **Relaxation Response** even if no
change is noticeable? _____

Yes. Decreased heart
rate always occurs and
this, by itself, is a
calming influence.

32
To begin the Relaxation Response sit comfortably in a place where no one is likely to disturb you. This could be in a classroom at the beginning of a test. Could you achieve this first step in a library? _____

Yes

33
While sitting comfortably in a quiet environment, try to insulate yourself from other external distractions. Would it be helpful to close your eyes? _____

Probably so. Most people find they are less likely to be distracted if their eyes are closed.

34
Relax the muscles in your body by starting with your toes and working slowly upwards. Try relaxing your toes, feet, ankles, knees, and thighs in that order. Write a sentence or two describing what happened.

_____

Most people who try this relaxing first become aware of a tenseness in their muscles and then, as they begin investigating this tenseness, they feel the muscles lengthen and soften. If this did not happen to you, keep trying because it will.

35
After you have relaxed the muscles in your body, begin to pay attention to your breathing. Try closing your eyes and concentrating on your breathing for twenty inhalations. Then write a sentence or two about what happened.

_____

For most people, when they begin paying attention to their breathing they feel some tightness in their chests and irregularities in their inhaling and exhaling. With continued attention, however, the chest tightness eases and the breathing becomes more regular.

36

Breathe through your nose and as you become
aware of your breathing pattern start
saying "one" every time you exhale. The
objective of repeating the word "one"
with each breath is to allow you to dwell
on the word rather than all the thoughts
which may run through your head. Does the
word "one" create any discordant thought
or arouse any emotions for you? _____

If your answer is "No" then
it will work fine for you. If
your answer is "Yes" then
another word like "out,"
"sack," or "all" needs to
be used.

37

Have you selected a word, like "one," to
say each time you exhale? _____

Hopefully the answer is
"Yes." If not, select
another word.

38

It's time to give the Relaxation Response
a test run. Begin by sitting very comfortably in a quiet place with your eyes
closed. As soon as you feel you have
tried to achieve this form of personal
insulation read the "answer" below.

Don't worry about this being
a difficult task because the
opposite is true. Just sit
quietly with your eyes
closed.

39

The next step is to relax your muscles.
This might take a while so don't try to
hurry. Begin with your toes and relax
each muscle you become aware of. As
soon as you feel you have tried to
achieve this muscle relaxation read the
"answer" below.

Again, don't worry about
this being difficult. Whatever relaxation was gained
this time is progress. This
effort becomes more effective
and easy with practice.

40
You should now be sitting comfortably with your body quite relaxed. The next step is to pay attention to your breathing. Breathe through your nose. As you pay attention to your breathing, it should become more regular. As soon as you have tried to achieve regular, easy breathing read the "answer" below.

The only suggestion you might need is this: Don't worry about how you are doing. **Just** try it and see what happens.

41
Maintain your comfortable, sitting position. As you breathe out (exhale) say the word "one." Repeat this for twenty breaths and then read the "answer" below.

You have just practiced the process which gets you into Benson's Relaxation Response. If the full benefits were desired you would maintain the situation for ten to twenty minutes.

ᴨᴨᴨᴨᴨᴨᴨᴨ   ᴨᴨᴨᴨᴨᴨᴨᴨ   ᴨᴨᴨᴨᴨᴨᴨᴨ   ᴨᴨᴨᴨᴨᴨᴨᴨ   ᴨᴨᴨᴨᴨᴨᴨ

THE FOUR STEPS OF THE RELAXATION RESPONSE

The four steps in the Relaxation Response are:
1) Sit quietly and comfortably with eyes closed.

2) Relax muscles. Begin with the feet and work up to the head.

3) Breathe through the nose. When breathing becomes regular, say "one" on each exhalation. Maintain this for ten to twenty minutes.

4) Let thoughts drift through the mind. Don't hang onto them and don't try to push them away. Pretend the thoughts are clouds which drift in and out of the environment.

42
Should you worry about doing the Relaxation
Response "correctly?" _____

No. The only way you can
do it incorrectly is to
worry about it.

43
Practicing the Relaxation Response will
improve the use of it. It will be easier
to get comfortable, relax muscles, breathe
regularly, and maintain a passive attitude
toward the thoughts that go through your
mind. Is it necessary to practice the
Relaxation Response before you achieve
its benefits? _____

No. The Relaxation
Response will be bene-
ficial each time the
process is used.
Practice makes it easier
and more effective.

44
Do not practice the Relaxation Response within
two hours after eating because the activity
of the digestive system interferes. If you
eat breakfast at 7 a.m. is it okay to practice
the Relaxation Response at 10 a.m.? _____

Yes. Allow two
hours after eating.

45
The Relaxation Response needs to be practiced
before you need it in an actual anxiety situation.
It is strongly suggested that you practice the
Relaxation Response two or three times each day
until you are familiar with the process. Should
you wait until you are in an anxiety situation
before practicing the Relaxation Response?

_____

No. Practice be-
fore you need it.

46
Before proceeding further you should spend
three days practicing the Relaxation
Response. What is the next assignment
you need to fulfill?

_____

Practice the Relaxation
Response for the next three
days. Do not continue until
you have practiced the
Relaxation Response.

CHAPTER 6   QUESTIONS

This is a book for math anxiety, but very little mathematics has been presented yet. This is because your defenses from further anxiety had to be developed and your strategies for handling the anxiety had to be learned. Now you are ready to begin to actually confront some mathematics, but the following confrontation allows **you** to remain in **control.**

One way in which you will maintain control is the design of this book. The book allows you to set the pace at which you will work. Self-pacing makes it possible for the reader to enjoy great success when the going is easy; it also allows the learner to go slowly and carefully when the material is difficult or threatening.

Another way in which you will maintain control is the fact that you now know the Relaxation Response. Whenever you need it, take a few minutes and use the Relaxation Response. Don't let the anxiety grow; keep it at a low level which will not interfere with your learning or understanding.

In the beginning, you may find that it is necessary to use the Relaxation Response frequently. This is natural and no cause for great concern because as you proceed from one successful experience to another you will find your interruptions for the Relaxation Response diminishing. This does not mean that you will completely overcome your anxiety. That would be most desirable, but unrealistic. It is likely that your anxiety will be reduced, but not eliminated. Remember, you are learning to reduce your anxiety to such an extent that it will no longer act as a barrier to other goals. Learning to cope with your anxiety is both a realistic and desirable objective. You are now at a point where you can experience real progress toward that objective.

On the next page, two figures are shown. Just look at the figures carefully and then answer the questions at the bottom of that page.

4

7

1. If you experienced any anxiety looking at the figures above, take a few minutes to practice the Relaxation Response. When you are calm, go to the next question.

2. Which of the figures above do you like best? Most people "like" some numbers better than others. Draw a happy face on the figure to show that it is pleased you like it best.

3. Draw a sad face on the figure you didn't like best. This will show that the figure is sad that you didn't like it best.

On the next page are two figures. Again, look carefully at the two figures and then answer the questions at the bottom of that page.

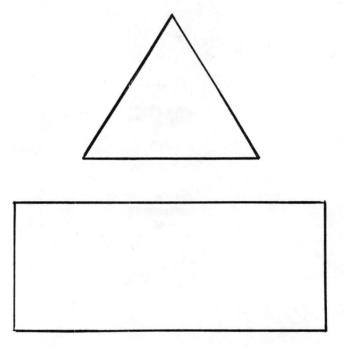

1. If you experienced any anxiety looking at the figures above, take a few minutes to practice the Relaxation Response. When you are calm, go to the next question.

2. Which of the figures do you like best? Draw a happy face on it.

3. Draw a sad face on the other figure.

On the next page are two more figures. Again, look carefully at the two figures and then answer the questions at the bottom of the page.

1. If you experienced any anxiety looking at the figures, take
   a few minutes to practice the Relaxation Response. When you
   are calm, go to the next question.

2. Which of the figures do you like best? Draw a happy face on
   it.

3. Draw a sad face on the other figure.

> On the next page are two figures. Again, look
> carefully at the two figures and then answer the
> questions at the bottom of the page.

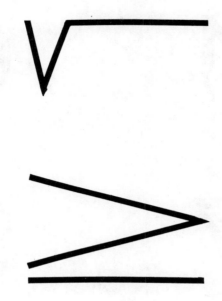

1. If you experienced any anxiety looking at the figures, take a few minutes to practice the Relaxation Response. When you are calm, go to the next question.

2. Which of the figures do you like best? Draw a happy face on it.

3. Draw a sad face on the other figure.

   On the next page are four figures. As before, look carefully at the figures and then answer the questions at the bottom of the page.

$$4 \qquad 6$$

$$7 \qquad 9$$

1. Sometimes, four numerals on a page will raise someone's anxiety while two numerals or figures does not have that effect. If your anxiety was raised by the numerals above, use the Relaxation Response. When you are calm, go to the next question.

2. Choose the numeral above that you like best. This gives the numeral a little personality. Draw a circle around this numeral.

   On the next page there are four more numerals. Look carefully at each figure and then answer the questions at the bottom of the page.

3   8

2   5

1.   If the numerals above raised your anxiety, use the Relaxation Response. When calm, go on to the next question.

2.   Choose the numeral above that you liked best. Draw a square around the numeral.

     On the next page are two math problems. **DO NOT DO THE PROBLEMS.** Look at them carefully and then answer the questions at the bottom of the page.

5 + 3 =

9 − 4 =

1. Sometimes, math problems raise anxiety while individual numerals do not have that effect. If these problems raised your anxiety, use the Relaxation Response. When calm, go to the next question.

2. Place a capital letter A on the answer blank of any problem that made you anxious.

3. Place a capital letter C on the answer blank of any problem you felt comfortable about.

4. If you think you know the answer to either problem write it down.

5. The answers to the problems are 8 and 5, respectively. Check the answer for each problem you worked.

On the next page are two more problems. DO NOT WORK THE PROBLEMS. Look at them and then answer the questions at the bottom of the page.

48 x 7 =

912 ÷ 6 =

1. Sometimes, multiplication and division problems raise more anxiety than addition or subtraction problems. If your anxiety was raised, use the Relaxation Response. When calm, go to the next question.

2. Place a capital letter A on the answer blank for any problem that raises your anxiety. Place a C on the answer blank of any problem that leaves you comfortable.

3. If you think you can work a problem correctly, do it now.

4. The answers to the problems above are 336 and 152. Check the answer for each problem you worked.

On the next page are two more problems. **DO NOT WORK THEM.** Look at them and then answer the questions at the bottom of the page. Do not start the next page immediately if you were inhibited by anxiety on this one.

$$\frac{5}{8} + \frac{3}{4} = \underline{\qquad}$$

$$\frac{2}{3} \times \frac{5}{8} = \underline{\qquad}$$

1. Fractions may be more threatening than whole numbers. If the problems above raise your anxiety, use the Relaxation Response. When calm, go to the next question.

2. Label any problem that raises your anxiety with an A and any problem that leaves you comfortable with a C.

3. If you know how to work any problem, do it now.

4. The answers to the problems are $\frac{11}{8} = 1\frac{3}{8}$ and $\frac{5}{12}$.

Check the answers for any problems you worked.

On the next page are two more problems. Your first task is to look carefully at the problems and see whether they raise your anxiety. Then answer the questions at the bottom of the page. Do not start the next page immediately if you were inhibited by anxiety on this one.

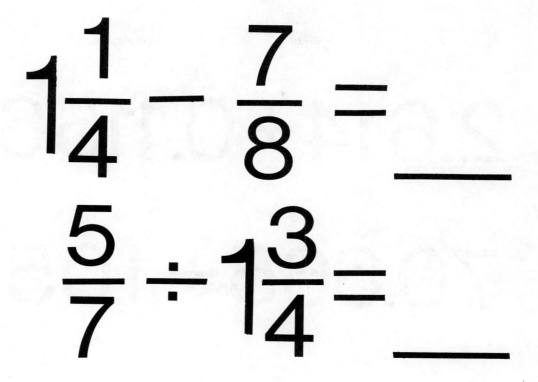

$$1\frac{1}{4} - \frac{7}{8} = \underline{\hspace{2cm}}$$

$$\frac{5}{7} \div 1\frac{3}{4} = \underline{\hspace{2cm}}$$

1. If the problems above raise any anxiety, use the Relaxation Response. When calm, go to the next question.

2. Label any problem that raises your anxiety with an A and any problem that leaves you comfortable with a C.

3. If you can work any problem, do it now.

4. The answers to the problems are $\frac{3}{8}$ and $\frac{20}{49}$.

Check the answer to any problems you worked.

Two more problems are shown on the next page. Look at the problems carefully and then answer the questions at the bottom of the page. Do not start the next page immediately if you were inhibited by anxiety on this one.

$2.614 + 0.1156$

$70.665 \div 1.05$

1. The problems above contain decimal numerals. If they cause any anxiety, use the Relaxation Response. When calm, go to the next question.

2. Label any problem that raises your anxiety with an A and any problem that leaves you comfortable with a C.

3. If you can work either problem, do it now.

4. The answers to the problems are 2.7296 and 67.3. Check the answer to any problem you worked.

       Two more problems are shown on the next page. Look at the problems carefully and then answer the questions at the bottom of the page. Do not start the next page immediately if you were inhibited by anxiety on this one.

$.47 \times 841.3 =$

$15.2 - 4.186 =$

1. Two more problems containing decimal numerals are shown above. If they cause any anxiety, use the Relaxation Response. When calm, go to the next question.

2. Label any problem that raises your anxiety with an A and any problem that leaves you comfortable with a C.

3. If you can work either problem, do it now.

4. The answers to the problems are 395.411 and 11.014. Check the answer to any problem you worked.

Do not start the next page immediately if you were inhibited by anxiety on this one.

6% of ___ is 912

___% of 4763 is 1524.16

1. The problems above involve percent. If the problems cause any anxiety, use the Relaxation Response. When calm, go to the next question.

2. Label any problem that raises your anxiety with an A and any problem that leaves you comfortable with a C.

3. If you can work either problem, do it now.

4. The answers to the problems are 15,200 and 32%. Check the answer to any problem you worked.

   The following exercise tests your reactions to a large number of problems. Do not begin it immediately if you were inhibited by anxiety on this page.

CHAPTER 6   EXERCISE

DIRECTIONS:

DO NOT DO ANY OF THESE PROBLEMS. Instead, look carefully at each problem. If a problem causes no anxiety or very little then mark it with a C for comfortable. If a problem causes you enough anxiety to interfere with your ability to do it correctly mark it with an A for anxious. If a problem is beyond your academic background mark it with a B for beyond.

1. $8 + 7 =$ _____

2. $9 - 7 =$ _____

3. $6 \times 7 =$ _____

4. $2,314 + 958 =$ _____

5. $36 \div 4 =$ _____

6. $15 + 8 =$ _____

7. $8,346 + 2,309 + 459 =$ _____

8. $8,906 \times 47 =$ _____

9.   Is 278 an even number? __

10. $5,000 - 2358 = -$ _____

11. 8058 x 498 = _____

12. Divide 5,679 by 38 and show the remainder. _____

13. 63,582 – 3,998 = _____

14. 57.93 x 0.78 = _____

15. Divide 47,893 by 46 and show the remainder. _____

16. Simplify $\frac{32}{50}$ _____

17. 2.706 x 4.08 = _____

18. Divide 57.36 by .93 and round off at two decimal places. _____

19. Which is greater $\frac{5}{7}$ or $\frac{2}{3}$? _____

20. $\frac{7}{8} + \frac{5}{8}$ = _____

21. Divide 0.0873 by 5.6 and round off at four decimal places. _____

22. Change $\frac{33}{8}$ to a mixed number. _____

23. $\frac{5}{9} + \frac{7}{12}$ = _____

24. $\frac{7}{8}$ x $\frac{4}{9}$ = _____

25. Change 5 $\frac{2}{7}$ to an improper fraction. _____

26. $\frac{5}{6} - \frac{2}{9}$ = _____

27. $3\frac{3}{7}$ x $\frac{5}{6}$ = _____

28. 67% of 3706 = _____

29. $2\frac{1}{2} - 1\frac{7}{9}$ = _____

30. $\frac{5}{7} \div \frac{2}{3}$ = _____

31. _____% of 1406 is 745.18

32. Find n when $\frac{57}{48} = \frac{12}{n}$ _____

33. $2\frac{3}{8} \div 6$ = _____

34. 14% of _____ is 82.04

35. If gasohol is mixed with two parts of alcohol to every nine parts of gasoline, how many gallons of gas should be mixed with 57 gallons of alcohol?

_____

36. Solve $3x - 7 = 22$ _____

37. If 5 of every 7 medical students are males, how many males would be in a medical school class of 518? _____

38. Simplify $3(2x - 5) - 2(7x - 1)$ _____

39. $6x^2y^5(x) - 4x^3y^5 =$ _____

40. When twice a number is increased by 5 the result is 23. Find the number. _____

CHAPTER 6   QUESTIONS

1.   Is there a particular type of numeral, number, symbol, or figure that is associated with your anxiety?

2.   Is there a particular type of problem (addition, subtraction, multiplication, division) that is associated with your anxiety?

3.   Does the number of problems you see on a page affect the level of your anxiety? If so, how might you handle that situation?

4.   Did you find that drawing a happy or sad face on a figure altered your feelings toward it? If so, how might you use that information?

CHAPTER 6   GROUP ACTIVITIES

1.   Conduct a group Relaxation Response with one of the members directing the group through each phase.

2.   Use crayons and colored magic markers to write sample problems for the group. What effects do colors have on anxiety?

Chapter 7

A Successful Math Learning Experience

Reducing or eliminating math anxiety is a difficult task, but many people have overcome feelings of helplessness. In almost all cases, however, three critical factors played a crucial role.

mmmmmm   mmmmmm   mmmmmm   mmmmmm   mmmmmm

THREE CRITICAL FACTORS IN OVERCOMING ANXIETY

These three factors are:

1) The individual was able to form a more realistic understanding of his/her situation.

2) The individual was pulled, forced, urged, or driven to try again even though the situation seemed similar to the hopeless one that created the anxiety.

3) The individual experienced repeated success on a series of tasks which previously seemed both onerous and hopelessly impossible.

Notice that all three of these factors allow the individual to regain control over situations that were previously believed to be exempt from any control.

In this book you have already received information and participated in activities specifically designed to completely fulfill the first factor named above. A realistic understanding of your own particular anxiety situation was acquired in the first five chapters of this book. The experiments with dogs, the autobiographical data you provided in words and pictures, and the information on learning, teaching, and study skills give you a new, and better, perspective on the math learning experiences of your past, present, and future.

Real progress has also been achieved toward completing the second factor listed above in overcoming your anxiety. The Relaxation Response gives you a strategy for coping with the physically and emotionally disabling aspects of your anxiety. The questions of Chapter 6 not only allowed you to practice the Relaxation Response, but also encouraged, forced, pulled, or drove you to begin interacting again with mathematics. This process will continue in these last four chapters of this book.

This leaves the third factor as the only healing element still to be achieved. That is the major purpose of the remainder of this book. It is designed to give you repeated success on a series of mathematical tasks.

"When I calmed down, it seemed to get easier."

To take full advantage of the following material, be certain that you always work through the three steps in the learning process:
1) awareness, (focusing)
2) action-response, and
3) feedback

The instruction you will be given is in the statement-answer form used earlier.

Each numbered statement-answer is a **frame** and represents a complete learning cycle. The left column of a frame gives you both an **awareness** and an opportunity for **action-response.** The frame answer is the **feedback.**

Work each frame as a complete lesson. Look for something new or different as the awareness step. Write the answer as your action-response. Check your answer immediately to receive your feedback.

Remember, the objective of the following material is to give you repeated success on a series of mathematical tasks. That is an essential element in reducing or eliminating math anxiety. If you are not achieving that repeated success:

a) Use the Relaxation Response before resuming work on the frames.

b) Make certain that you do each of the three steps of learning on each frame.

c) Practice the four needed study skills for mathematics.

d) Do not continue unless you are consistently getting correct answers.

UNIT 1:   SOLVING WORD PROBLEMS REQUIRING ADDITION

******* **Example** ******* ******* ******

VERTICAL ADDITION REQUIRING "CARRYING"

To add 4093 and 2398 the numerals are arranged
vertically so that **digits with the same
place value are in the same column.**

```
  4 0 9 3
  2 3 9 8
  6 4 9 1
```

Each column from right to left is separately
added.   Whenever the answer to any column
exceeds nine, the process of carrying is used.

The "carrying" marks from the additions of the
ones and the tens columns are shown in the
figure at the right.   Use these "carrying"
marks if they improve your accuracy.

```
  1 1
  4 0 9 3
  2 3 9 8
  6 4 9 1
```

ⁿⁿⁿⁿⁿⁿⁿⁿ ⁿⁿⁿⁿⁿⁿⁿⁿ ⁿⁿⁿⁿⁿⁿⁿ ⁿⁿⁿⁿⁿⁿⁿ ⁿⁿⁿⁿⁿⁿ

1
Arrange the three addends 635, 43, and 8136 in
columns and find the sum of only the right-hand
column. _____

```
      1
    6 3 5
      4 3
  8 1 3 6
        4
```

2
Arrange the three addends 4315, 216, and 483 for
column addition and find the sum of only the
right-hand column. _____

```
      1
  4 3 1 5
    2 1 6
    4 8 3
        4
```

31
Add 905 + 1483 + 384 using column addition and
showing all carrying. _____

```
  1 1 1
    9 0 5
  1 4 8 3
    3 8 4
  2 7 7 2
```

4
Add 674 + 8905 + 655 using column addition and
showing all carrying. _____

```
    2 1 1
      6 7 4
    8 9 0 5
      6 5 5
  1 0 2 3 4
```

5

$426 + 5270 + 8654 =$ _____   14,350

6

$7179 + 869 + 7658 =$ _____   15,706

7

$6845 + 97403 + 6510 =$ _____   110,758

8

$4,326,913 + 58,496 =$ _____   4,385,409

9

Add $67,315 + 412,684 + 5,094 + 678.$ _____   485,771

10

Add $575,903 + 96,318 + 4,867 + 718 + 987,865.$ _____   1,665,671

*******   **Example**   *******   *******   ******

ADDITION TERMS

The numbers in the addition problem at the right
are **addends.** 5469, 395, and 4067 are **addends.**

$$\begin{array}{r} 5469 \\ 395 \\ +4067 \\ \hline 9931 \end{array}$$

The answer of an addition problem is its **sum.**
9931 is the **sum** of the problem at the right.

11

The answer to an addition problem is its **sum.** The
word **total** is also used for the answer to an
addition problem. Find the **total** of 87,306 and
422,312. _____   509,618

111

12

A rancher has three pastures measuring 4,352
acres, 518 acres, and 47 acres. To find his
total pasture land the numbers should be
_____.   added

13

Sam Jones owes two bills of $512 and $87. To
find the **sum** of money owed the numbers should
be _____.   added

14

A druggist prepared 45 prescriptions on Monday, 32 on Tuesday, and 54 Wednesday. What
was the **total** number of prescriptions those
three days? _____   45 + 32 + 54 = 131

15

A nurse worked 20 days in June, 24 days in
July, 19 days in August, and 12 days in
September. What was the **sum** of days
worked those four months? _____   20 + 24 + 19 + 12 = 75

NEW IDEA

16

It makes no sense to add 3 cats and 4 dogs
because the numbers have different labels
- cats and dogs. Is it sensible to add 57
apples and 12 peaches? ____   No

17

Is it sensible to find the **sum** of 47 Fords
and 53 Dodges? _____   No

18

To add two numbers they must have the same
label. Is it possible to find the **total** of
19 cows and 483 cows? _____   Yes, 502 cows

Post Quiz #1

This quiz reviews the preceding unit. Answers are at the back of the book.
Do not continue until all problems are understood.

1
   374 + 2331 + 47

2
      10,003 + 412 + 78 + 39 + 112

3
   2398 + 97 + 783 + 79

4
      18 + 4374 + 399 + 415 + 10,473

5
The words _____ and _____ mean the answer to an addition problem.

6
Bill loaded the following amounts of concrete onto a truck: 100 lbs, 50 lbs,
25 lbs, 93 lbs. Find the total.

7
Find the sum of Mel Black's investments if he has separately invested
$46,724 and $9,764.

8
Numbers should not be added if they have _____ (the same, different) labels.

‍‍‍‍‍‍‍‍‍‍‍‍‍‍‍   ‍‍‍‍‍‍‍‍‍‍   ‍‍‍‍‍‍‍‍‍‍   ‍‍‍‍‍‍‍‍‍‍   ‍‍‍‍‍‍‍‍

UNIT 2: WORD PROBLEMS REQUIRING SUBTRACTION

*******   **Example**   *******   *******   ******

MINUEND AND SUBTRAHEND

The numbers in a subtraction problem are the
minuend and **subtrahend**. The answer is
the **difference**.

In the problem at the right the **minuend**
is 5867; the **subtrahend** is 2514; and the
difference is 3353.

5867	**minuend**
− 2514	**subtrahend**
3353	**difference**

1
The numbers in a subtraction
problem are the **minuend** 4653 minuend
and **subtrahend.**   -3197 subtrahend

In the problem above, the **minuend** is 4653 and the
subtrahend is _____ .   3197

2
The **subtrahend** of the problem
at the right is 3197 and the   4653
minuend is _____ .   -3197   4653

3
The **minuend** of 5 - 3 is 5 and the **subtrahend** is 3.
What is the **minuend** of 56 - 47?_____   56

4
What is the **subtrahend** of 963 - 673? _____   673

5
The **minuend** of 6715 - 2438 is _____ .   6715

6
The **subtrahend** of 4693 - 1946 is _____ .   1946

NEW IDEA

7
In an addition problem the numbers being added are
called **addends.** In a subtraction problem the
numbers are the _____ and _____ .   minuend,
                   subtrahend

8
An addition problem has two or more addends. Does
a subtraction problem have two or more minuends?
_____   No, only one

9
In the problem 2 + 4 - 6, 4 is a/an _____ .   addend

10

The answer to an addition problem is called its **sum.**
What is the sum of 476 + 51? _____   527

NEW IDEA

11

The answer to a subtraction problem is called its
difference.   What is the **difference** of
57 - 41 = 16? _____   16

12

In the subtraction problem 12 - 5 = 7 the number 7
is called the _____.   difference

13

Use the numbers from the problem 83 - 32 = 51.
Find the **sum** of the **subtrahend** and the
difference. _____   32 + 51 = 83

14

At the right are shown a   459 - 127 = 332
subtraction and a related
addition problem.   127 + 332 = 459

The sum of the addition problem is the _____ of
the subtraction problem.   minuend

NEW IDEA

15

To check a subtraction problem, find the sum of the
subtrahend and the difference.

To check 9 - 2 = 7, add _____ and _____.   2 and 7

16

To check a subtraction, add the subtrahend and the
difference to see if it equals the _____.   minuend

115

17
Does the subtraction below check? _____
    16 - 9 = 5 and 9 + 5 = 14.

No, 14 is not
the minuend

18
Subtract and check.   573 - 268 = _____

573 - 268 = 305
268 + 305 = 573

19
Subtract and check.   917 - 432 = _____

917 - 432 = 485
432 + 485 = 917

20
Subtract and check.   465 - 199 = _____

465 - 199 = 266
199 + 266 = 465

21
Subtract and check.   600 - 362 = _____

600 - 362 = 238
362 + 238 = 600

NEW IDEA

22
Subtraction is used to solve problems which ask "how
much more" or "how many more."   Would subtraction
be used to solve: If Sara has $15 and Mike has $38,
how many more dollars does Mike have? _____

Yes, 38 - 15 = 23

23
If Dan weighs 158 pounds and Joe weighs 141 pounds,
how much more does Dan weigh?
Would subtraction be used to solve? _____

Yes, 158 - 141 = 17

24
Subtraction is used to solve problems that ask "how
much was left." Peter took 78 cents to the store
and bought a 43 cent pen. How much money was left?
Would subtraction be used to solve? _____

Yes, 78 - 43 = 35

25
Would subtraction be used to solve? _____
Farmer Smith sold 930 acres from his 2536 acre
ranch. How many acres were left?

Yes,
2536 - 930 = 1606

26
In an addition problem should the numbers have the
same label? _____

Yes

27
In both addition and subtraction problems the
numbers must have the same labels. Does it make
sense to add 5 dogs and 3 cats? _____

No, cats and dogs
are different labels.

28
Does it make sense to subtract 47 cows from 58 pigs?
_____

No, cows and pigs
are different labels.

29
Does it make sense to subtract 97 boxes from 482
boxes? _____

Yes, both labels are
the same.

30
If Sam worked 25 days in January and 18 days in
February, how many more days did he work in January?
_____

25 - 18
7 days

31
Paul had 33 vacation days coming. After a 15 day
vacation, how many more days were left? _____

33 - 15
18 days

32
A sculptor started a statue from a block of granite
weighing 942 pounds. He estimated chipping off 150
pounds. How much granite was left?

942 - 150
792 pounds

Chapter 7

Post Quiz #2

This quiz reviews the preceding unit.   Answers are at the back of the book.
Do not continue until all problems are understood.

1
In the problem 17 - 9 = 8 the minuend is _____ .

2
In the problem 47 - 35 = 12 the difference is _____ .

3
In the problem 56 - 11 = 45 the subtrahend is _____ .

4
Solve and check.

a)   5614 - 2163

b)   40,936 - 15,487

c)   6149 - 2847

d)   50,000 - 19,306

5
If a nurse earns $147 a week and a cab driver earns $205 a week, how much
more does a cab driver earn each week?

6
If a man has read 387 pages of a 520-page book, how many pages are left?

UNIT 3:   WORD PROBLEMS REQUIRING MULTIPLICATION

******* **Example** ******* ******* ******

MULTIPLICATION TERMINOLOGY AND APPLICATIONS

In a multiplication problem the numbers being multiplied are
factors.   For example, in the problem 6 x 9 = 54, the numbers
6 and 9 are **factors.**   54 is the **product.**

438	**multiplicand**
x 69	**multiplier**
3942	**partial product**
2628	**partial product**
30222	**product**

When two numbers are multiplied
using a vertical arrangement of
the **factors,** the **factor**
on top is the **multiplicand**
and the **factor** on the bottom
is the **multiplier.**

1

The numbers in a multiplication problem are called
factors. In the problem 6 x 7 = 42 the **factors**
are _____ and _____.   6, 7

2

In the problem 8 x 7 = 56, the **product** is 56 and the
factors are _____ and _____.   7, 8

3

What are the **factors** in 3 x 6 = 18? _____   3, 6

4

In the problem 10 x 73 = 730, the product is 730 and
the factors are _____ and _____.   10, 73

NEW IDEA

5

When one factor of a multiplication problem is 10,
the product will always be the other factor with
a zero affixed.

            10 x 98 = 980
            10 x 45 = _____   450

6

            10 x 543 = 5430
            10 x 764 = _____   7640

NEW IDEA

7

There is a shortcut for multiplying by 100. Affix
two zeros to the other factor.

            100 x 47 = 4700
            100 x 83 = _____   8300

8
There is a shortcut for multiplying by 1000. Affix
three zeros to the other factor.

        1000 x 675 = 675000
        1000 x 432 = _____   432000

NEW IDEA

9
Word problems requiring multiplication frequently
contain the word **"each."** If the word **"each"** appears
in a word problem then it probably indicates that
the numbers should be _____.   multiplied

10
What word often indicates multiplication in a
word problem? _____   each

11
If candy bars cost 15 cents **each,** then the price of
7 bars is found by _____ 15 and 7.   multiplying

12
What word in the phrase, "Six boys, each weighing 74
pounds," indicates multiplication may be needed? _____   each

13
Six boys, each weighing 74 pounds, weigh a total
of _____ pounds.   444 pounds

14
If an orderly works 8 days and each day works 7
hours, how many hours were worked? _____   56 hours

NEW IDEA

15
Numbers in a multiplication problem do not have to
have the same label. Do numbers in an addition
problem have to have the same label? _____   Yes

16

Numbers in a multiplication problem may have different labels. Can numbers in a subtraction problem have different labels? _____

No

17

A therapist needed to spend 14 hours with each of 13 patients. Find the total number of hours needed. _____

182 hours

18

A shipment of sports cars consists of 27 units priced at $9,700 each. What is the value of the shipment? _____

$261900

19

A geological survey team brought back 548 specimens that weighed an average of 29 kilograms each. Find the total weight of the specimens. _____

15892 kilograms

20

A bottle of medicine contains 83 pills. Each pill weighs 27 milligrams. What is the weight of the pills? _____

2241 milligrams

21

If each ticket in a bundle of 2,000 tickets costs 76 cents, what is the cost of the bundle? _____

152000 cents

22

A case holds 64 bottles. If each bottle contains 55 grams of mercury, how much mercury is in the case? _____

3520 gm

Chapter 7

Post Quiz #3

This quiz reviews the preceding unit. Answers are at the back of the book.
Do not continue until all problems are understood.

1
The numbers in a multiplication problem are called the multiplier and
the multiplicand. Another word for either number in a multiplication
problem is _____.

2   Use a multiplication shortcut to find the answers for:

     a)   10 x 56 = _____   b)   10 x 5678 = _____

     c)   100 x 61 = _____   d)   100 x 48 = _____

3
Each test tube in a box weighs 57 grams. Find the total weight of the test
tubes if there are 144 tubes in the box.

4
Seventeen girl scouts went on a camping trip. Each scout carried a 26 pound
backpack. What was the total weight in the backpacks?

    ᴍᴍᴍᴍᴍ   ᴍᴍᴍᴍᴍ   ᴍᴍᴍᴍᴍ   ᴍᴍᴍᴍᴍ   ᴍᴍᴍᴍᴍ

UNIT 4:   A REVIEW OF LONG DIVISION

       ******   Example   *******   *******   ******

MAINTAIN VERTICAL COLUMNS CAREFULLY

In dividing 2741 by 6, it is important to
remember that one digit of the answer goes
directly above each digit of 2741.

In practice the first digit of the answer
that is written is not zero. This common
practice is shown in the example at the
right.

```
      456 R5
6 )2741
   -24
     34
    -30
     41
    -36
      5
```

1

Find the first digit of the answer for:

```
      _____
57 )25738
```

```
          4
      _____
57 )25738
   -228
     29
```

2

Find the first digit of the answer for:

```
      _____
83 )47362
```

```
          5
      _____
83 )47362
   -415
     58
```

3

Find the first digit of the answer for:

```
      _____
65 )19836
```

```
          3
      _____
65 )19836
   -195
      3
```

******* Example ******* ******* ******

THE LONG DIVISION PROCESS

The long division shown at the right follows exactly the same process as the previous examples.

To maintain the process, remember:

    (1) A digit of the answer will appear directly above each digit of 437,628 with the understanding that the zeros over 4 and 3 will not be written.

    (2) Maintain the columns carefully.

```
                1 2 6 1 R61
            _____
3 4 7 )4 3 7,6 2 8
      -3 4 7
        9 0 6
       -6 9 4
        2 1 2 2
       -2 0 8 2
          4 0 8
         -3 4 7
            6 1
```

4
Complete the following.

```
        26                    267 R17
35 )9362              35 )9362
   -70                   -70
    236                   236
   -210                  -210
    262                   262
   ————                  -245
                           17
```

5
Complete the following.
Caution: There is a zero
in the answer. Be
certain to place it
correctly.

```
58 )17635              304 R3
                  58 )17635
                     -174
                       23
                      -0
                      235
                     -232
                        3
```

6
Complete the following.

```
92 )40635             441 R63
                  92 )40635
                     -368
                      383
                     -368
                      155
                      -92
                       63
```

7
Complete the following.

```
87 )44134             507 R25
                  87 )44134
                     -435
                       63
                      -0
                      634
                     -609
                       25
```

8
Complete the following.

$$84\overline{)37605}$$

$$\begin{array}{r} 447 \text{ R}57 \\ 84\overline{)37605} \\ -336 \\ \hline 400 \\ -336 \\ \hline 645 \\ -588 \\ \hline 57 \end{array}$$

Post Quiz #4

This quiz reviews the preceding unit. Answers are at the back of the book. Do not continue until all problems are understood.

Divide.

1
$$65\overline{)8143}$$

2
$$59\overline{)29367}$$

3
$$43\overline{)9214}$$

4
$$28\overline{)14632}$$

5
$$17\overline{)5736}$$

6
$$85\overline{)9437}$$

7
$$72\overline{)14314}$$

8
$$96\overline{)25438}$$

9
$$15\overline{)9145}$$

10
$$71\overline{)6308}$$

11
$$48\overline{)90504}$$

12
$$76\overline{)14098}$$

UNIT 5:   WORD PROBLEMS REQUIRING DIVISION

****** **Example** ******* ******* ******

DIVISION NOMENCLATURE AND APPLICATIONS

In a division problem each of the numbers has a special name.   In the problem at the right, 628 is the **divisor,** 85 is the **quotient,** 53,672 is the **dividend,** and 292 is the **remainder.**

```
                8 5  R292
     6 2 8 )5 3,6 7 2
           -5 0 2 4
             3 4 3 2
            -3 1 4 0
               2 9 2
```

1

In the problem at the right 55 is the **divisor** and 948 is the **dividend.**

```
     55 )948
```

In the problem at the right the **divisor** is _____ .

```
     78 )9136
```

78

2

In the problem at the right the **dividend** is _____ .

```
     724 )41632
```

41632

3

In the problem at the right

```
         55 R11
     35 )1936
        -175
         186
        -175
          11
```

the **divisor** is 35, the **dividend** is 1936, the **quotient** is 55, and the **remainder** is _____ .

11

4

In the problem
at the right

$$\begin{array}{r} 54 \\ 14\overline{)756} \\ -70 \\ \hline 56 \\ -56 \\ \hline 0 \end{array}$$

the quotient is _____.

54

5

Find the divisor of

$$\begin{array}{r} 15 \text{ R3} \\ 6\overline{)93} \end{array}$$

_____

6

6

Find the dividend of

$$\begin{array}{r} 54 \text{ R4} \\ 8\overline{)436} \end{array}$$

_____

436

7

Find the quotient of

$$\begin{array}{r} 62 \\ 7\overline{)434} \end{array}$$

_____

62

NEW IDEA

8

To check the accuracy of a division answer multiply the quotient and the divisor. This product is added to the remainder. If the sum equals the dividend the division is correct.

Check:

$$\begin{array}{r} 14 \text{ R3} \\ 5\overline{)73} \end{array}$$

_____

$5 \times 14 + 3 = 73$

9

Show how to check

_____

$$\begin{array}{r} 4 \text{ R15} \\ 45\overline{)195} \end{array}$$

_____

$4 \times 45 + 15 = 195$

10
In a division problem with remainder zero, the
product of the _____ and _____ must equal
the _____.

quotient, divisor
dividend

NEW IDEA

11
To find the **average** of the numbers 59, 68, and 62
add the numbers and divide by 3.

Find the **average** of: 59, 68, and 62.

$$
\begin{array}{r}
63 \\
3\overline{)189} \\
-18 \\
\hline
09 \\
-9 \\
\hline
0
\end{array}
$$

12
Find the **average** of 257, 364, 295, and 308 by adding
the numbers and dividing by 4. _____

$$
\begin{array}{r}
306 \\
4\overline{)1224}
\end{array}
$$

13
Find the **average** of 19, 25, 30, 36, 25, and 27 by
counting the numbers and dividing their sum by
that count. _____

$$
\begin{array}{r}
27 \\
6\overline{)162}
\end{array}
$$

14
Find the average of 261 and 313. _____

$$
\begin{array}{r}
287 \\
2\overline{)574}
\end{array}
$$

15
Find the average of 679, 513, and 521. _____

$$
\begin{array}{r}
571 \\
3\overline{)1713}
\end{array}
$$

NEW IDEA

16
When a word problem uses the word **"each"** it
usually indicates a multiplication or a division.
Multiplication or division is often indicated by
the word, "_____."

each

17
A word problem with the word **"each"** usually
indicates multiplication or _____.

division

18
Solve the problem: If 6 men earn a total of $864,
what is the average earned by **each** man? _____
Hint: Divide 864 into 6 equal parts.

$$\frac{\$144}{6\,)\overline{864}}$$

19
Solve the problem: If a chemist has 845 milligrams
of medicine in 5 test tubes, what is the average in
each tube? _____
Hint: Divide 845 into 5 equal parts.

$$\frac{169}{5\,)\overline{845}}\ \text{mg}$$

20
If 5 nurses have a total of 65 vacation days, what
is the average vacation due each nurse? _____
Hint: Divide 65 into 5 equal parts.

$$\frac{13}{5\,)\overline{65}}\ \text{days}$$

21
If 8624 fluid ounces is to be equally distributed
into 7 containers, what amount should be put in
each container? _____

$$\frac{1232}{7\,)\overline{8624}}\ \text{fl oz}$$

22
Each tire in a sale is priced at $37. The total
value of the tires is $2,775. Find the number of
tires. _____
Hint: The answer will be the number of 37's
there are in 2775.

$$\frac{75}{37\,)\overline{2775}}$$

23
A total of 7101 grams of gold is to be made into
ornaments that weigh 9 grams each. How many
ornaments can be made? _____

$$\frac{789}{9\,)\overline{7101}}$$

24

If 893 pearls are to be made into necklaces
containing 19 pearls each, how many necklaces
can be obtained? _____

$$\begin{array}{r} 47 \\ 19\,\overline{)893} \end{array}$$

25

How many statues weighing 23 tons each can
be poured with 2,116 tons of concrete? _____

$$\begin{array}{r} 92 \\ 23\,\overline{)2116} \end{array}$$

Post Quiz #5

This quiz reviews the preceding unit.   Answers are at the back of the book.
Do not continue until all problems are understood.

1

In the problem at the right,
    the divisor is _____,
    the dividend is _____,
    the quotient is _____,
    and the remainder is _____.

$$\begin{array}{r} 19 \text{ R}38 \\ 46\,\overline{)912} \\ -46 \\ \hline 452 \\ -414 \\ \hline 38 \end{array}$$

2

Show how to check the division problem
at the right.

$$\begin{array}{r} 507 \text{ R}17 \\ 18\,\overline{)9143} \end{array}$$

3

Find the average of 57, 46, 52, and 49.

4

If 16 employees have total wages of $5712, find the average pay
of each employee.

5

    9378 cards are to be separated into 6 equal piles.   How many
cards will be in each pile?

CHAPTER 7 POST-TEST

This test reviews the objectives of the chapter. The student is expected
to know how to do **all** of these problems before finishing the chapter.
Answers for this test are at the end of the book.

1
A druggist filled 432 prescriptions
in July and 856 in August. What
was the sum for the two months?

2
Hospitals in Town A had 913 beds and
in Town B 476 beds. What was the
total number of beds in the two towns?

3
The distance from Town A to Town B is 435 miles and from Town B
to Town C is 658 miles. Find the sum of the distances.

4
A carpenter earned $319 one week and $67 the next week. What were
the total earnings during the two week period?

5
School A had 825 sophomores this year and School B had 573. How
many more sophomores did School A have than school B?

6
A manufacturer delivered 1358 orders in April and 902 in May. What
was the difference in orders delivered for the two month period?

7
It is 486 miles from Town A to Town B and only 195 miles from Town A to
Town C. How much further is it from Town A to Town B than from A to C?

8
Farmer A has 853 acres and Farmer B has 290 acres. What is the difference
in the size of their farms?

9
A grocer received 34 cartons of peas with 18 cans in each carton.
How many cans of peas were there?

10
Jane Smith pays $185 each month for rent. How much rent will she
pay in 12 months?

11
The numbers of a multiplication problem are the multiplicand and the
multiplier or both may be called _____.

12
The answer to a multiplication problem is its _____.

13
A druggist received 28 cartons with 24 jars of vitamins in each carton.
How many vitamin jars were there?

14
Bob Jones is paid $204 each week.   How much does he earn in 15 weeks?

15
In the problem at the right
the divisor is _____,
the dividend is _____,
and the quotient is _____.

$$\begin{array}{r} 58 \\ 47\ \overline{)2726} \end{array}$$

16

$$65\ \overline{)13296}$$

17
Find the average of 57, 65,
72, and 66.

18
480 fluid ounces of saline solution are to be distributed equally into 15
containers.   How many fluid ounces should be poured into each container?

19
In the problem at the right,
the divisor is _____,
the dividend is _____,
and the quotient is _____.

$$\begin{array}{r} 65 \\ 46\ \overline{)2990} \end{array}$$

20 $4\ \overline{)3615}$ 21 $7\ \overline{)6213}$ 22 $8\ \overline{)1946}$ 23 $5\ \overline{)6053}$

24 $28\ \overline{)5213}$ 25 $37\ \overline{)36541}$ 26 $63\ \overline{)65314}$ 27 $58\ \overline{)91326}$

28
Find the average of 52, 60, 67, and 61.

29
640 customers are to be equally distributed to 16 salespersons.
How many customers will each salesperson receive?

Chapter 8

Maintaining Success

You have started to change many long-term attitudes and habits that made learning mathematics both difficult and unlikeable. But, your complete recovery may take months or years. Old habits and feelings are difficult to extinguish and will return unless there is a conscious, persistent effort to develop new ones.

In this chapter, you need to stay constantly on guard against the return of old, counter-productive behaviors. It is extremely important, now that we are making progress, that your newly acquired study skills be cemented in place by consistent drill and practice.

ꓵꓵꓵꓵꓵꓵꓵ   ꓵꓵꓵꓵꓵꓵꓵ   ꓵꓵꓵꓵꓵꓵꓵ   ꓵꓵꓵꓵꓵꓵꓵ   ꓵꓵꓵꓵꓵꓵ

THE PURPOSE OF THIS SECTION

Two major objectives are to be achieved here. They are:

1) Continued practice with:
    a) the Relaxation Response.
    b) the three steps of the learning process.
    c) the four study skills of mathematics.

2) Repeated success with mathematical tasks.

Remember, this material is written for you to be successful. Enjoy that repeated success; it is an essential part of overcoming anxiety. Practice the new skills and information you have that need to become automatic responses for you. The more control you can exercise over your comfort level and the learning of mathematical tasks, the more complete will be this process of anxiety reduction.

UNIT 1: WORD PROBLEMS REQUIRING ADDITION OR SUBTRACTION OF FRACTIONS

******* **Example** *******   *******   ******

ADDING MIXED NUMBERS

To add two mixed numbers, like $3\frac{1}{5}$ and $5\frac{2}{3}$,

1) Add the two proper fractions.

2) Add the two whole numbers.

$$3 \quad \frac{1}{5} = 3 \quad \frac{3}{15}$$
$$5 \quad \frac{2}{3} = 5 \quad \frac{10}{15}$$
$$\overline{ 8 \quad \frac{13}{15}}$$

******* **Example** *******   *******   ******

SUBTRACTING MIXED NUMBERS

To subtract two mixed numbers which have different denominators, a common denominator must first be found.

The steps in subtracting $5\frac{3}{5}$ from $8\frac{1}{4}$ are shown below.

$$8 \frac{1}{4} = \quad 8 \quad \frac{5}{20} = \quad 7 \quad \frac{25}{20}$$
$$- 5 \frac{3}{5} = - 5 \quad \frac{12}{20} = - 5 \quad \frac{12}{20}$$
$$\overline{ 2 \quad \frac{13}{20}}$$

****** **Example** ****** ****** *****

TERMINOLOGY OF ADDITION OR SUBTRACTION OF FRACTIONS

The numbers in an addition problem are called **addends**. The answer to an addition problem is its **sum**.

$$4\frac{3}{8} \quad \text{addend}$$
$$+\ 1\frac{1}{8} \quad \text{addend}$$
$$5\frac{4}{8} = 5\frac{1}{2} \quad \text{sum}$$

The numbers in a subtraction problem are the **minuend** and **subtrahend**. The answer to a subtraction problem is the **difference**.

$$6\frac{7}{12} \quad \text{minuend}$$
$$-\ 5\frac{4}{12} \quad \text{subtrahend}$$
$$1\frac{3}{12} = 1\frac{1}{4} \quad \text{difference}$$

1

The answer to an addition problem is its **sum**.

Find the sum of $2\frac{1}{5}$ and $4\frac{3}{5}$. _____

$2\frac{1}{5} + 4\frac{3}{5} = 6\frac{4}{5}$

2

The answer to an addition problem is its_____

sum

3

The answer to a subtraction problem is its **difference**. Find the difference between $6\frac{7}{11}$ and $3\frac{2}{11}$. _____

$6\frac{7}{11} - 3\frac{2}{11} = 3\frac{5}{11}$

4

The answer to a subtraction problem is its _____.

difference

5
Find the sum of $3\frac{5}{8}$ and $2\frac{1}{4}$. _____

$$3\frac{5}{8} + 2\frac{2}{8} = 5\frac{7}{8}$$

6
Find the difference of $5\frac{9}{10}$ and $1\frac{1}{2}$. _____

$$5\frac{9}{10} - 1\frac{5}{10} = 4\frac{4}{10}$$
$$= 4\frac{2}{5}$$

7
Find the sum of $6\frac{8}{11}$ and $4\frac{1}{2}$. _____

$$6\frac{16}{22} + 4\frac{11}{22} = 10\frac{27}{22}$$
$$= 11\frac{5}{22}$$

8
Find the difference of $9\frac{1}{6}$ and $1\frac{3}{4}$. _____

$$8\frac{14}{12} - 1\frac{9}{12} = 7\frac{5}{12}$$

9
What is the sum of $\frac{3}{4}$ lbs of sugar and $1\frac{1}{2}$ lbs
of sugar? _____

$2\frac{1}{4}$ lbs of sugar

10
Does it make sense to add $4\frac{1}{2}$ lbs of flour and
3 lbs of sugar? _____

No, different labels

11
Does it make sense to add $4\frac{1}{2}$ ounces of flour and
$3\frac{1}{4}$ lbs of flour? _____

No, different labels

NEW IDEA

12
The words "how much more" usually indicate
subtraction. How much more is $2\frac{3}{4}$ yards than
$1\frac{5}{8}$ yards? _____

$$2\frac{6}{8} - 1\frac{5}{8} = 1\frac{1}{8} \text{ yds}$$

13
How much more is $7\frac{1}{2}$ days than $1\frac{2}{3}$ days? _____

$6\frac{9}{6} - 1\frac{4}{6} = 5\frac{5}{6}$ days

14
One capsule holds $20\frac{3}{10}$ grams while another holds

$15\frac{7}{10}$ grams. How much more is in the bigger
capsule? _____

$19\frac{13}{10} - 15\frac{7}{10} = 4\frac{3}{5}$

15
Ben bought two steaks. One weighed $2\frac{1}{4}$ lbs and
the other $1\frac{1}{2}$ lbs. How much steak did Ben
buy? _____

$2\frac{1}{4} + 1\frac{2}{4} = 3\frac{3}{4}$ lbs

16
One bottle held $4\frac{3}{8}$ liters of alcohol and another
held $1\frac{5}{8}$ liters. How much more did the first
bottle hold? _____

$3\frac{11}{8} - 1\frac{5}{8} = 2\frac{3}{4}$ liters

17
A specimen of iron ore weighed $4\frac{7}{10}$ kilograms and
another weighed $3\frac{1}{10}$ kilograms. What was the sum
of the weights of the two specimens? _____

$4\frac{7}{10} + 3\frac{1}{10} = 7\frac{4}{5}$ kgm

18
Mary worked $2\frac{5}{6}$ hours and Sally worked $7\frac{1}{4}$
hours. How much more did Sally work? _____

$6\frac{15}{12} - 2\frac{10}{12} = 4\frac{5}{12}$ hrs

19
Container A held $5\frac{1}{2}$ cups of milk while container
B held only $2\frac{3}{4}$ cups. Find the difference in the
amounts of milk the two containers can hold. _____

$4\frac{6}{4} - 2\frac{3}{4} = 2\frac{3}{4}$ cups

20

The price of a stock was 10\frac{1}{8}$ on Thursday and 8\frac{3}{4}$ on Friday. What was the difference in the price those two days? _____

$$9\frac{9}{8} - 8\frac{6}{8} = \$1\frac{3}{8}$$

21

Bill ran $1\frac{1}{4}$ hours in the morning and $2\frac{3}{5}$ hours in the afternoon. How many hours did he run altogether? _____

$$3\frac{17}{20} \text{ hours}$$

Post Quiz #1

This quiz reviews the preceding unit. Answers are at the back of the book. Do not continue until all problems are understood.

1

The answer to a subtraction problem is its _____.

2

The answer to an addition problem is its _____.

3

How much more is $9\frac{1}{4}$ inches than $4\frac{7}{8}$ inches?

4

Sarah bought two pieces of silk. One was $1\frac{2}{3}$ yards long and the other was $1\frac{3}{4}$ yards long. What was the total length of the silk?

5

One stock sells for 11\frac{3}{8}$ while another sells for 9\frac{1}{4}$. Find the sum of the prices of the stocks.

6

A block of iron weighed $12\frac{7}{10}$ kilograms while a block of gold weighed $2\frac{9}{10}$ kilograms. How much more did the iron weigh than the gold?

7

One capsule weighed $5\frac{1}{10}$ grams. Another weighed $3\frac{3}{5}$ grams. Find the difference in their weights.

8

If $2\frac{73}{100}$ meters of rope is cut from a piece $6\frac{3}{5}$ meters long, how much will be left?

UNIT 2: WORD PROBLEMS REQUIRING
MULTIPLICATION OR DIVISION OF FRACTIONS

*******   **Example**   *******   *******   ******

CANCELLING — A VALUABLE SHORTCUT IN MULTIPLYING FRACTIONS

One of the most valuable, useful shortcuts in arithmetic is the cancelling that is often possible when multiplying fractions.

Cancelling is possible when a numerator and a denominator have a common divisor greater than one. In the example at the right, 5 is a common factor. When the 5's are cancelled the multiplication problem becomes a simpler multiplication.

$$\frac{3}{5} \times \frac{5}{7} = \frac{3}{\overset{1}{\cancel{5}}} \times \frac{\overset{1}{\cancel{5}}}{7}$$

$$\frac{5}{5} = \frac{1}{1}$$

$$\frac{3}{1} \times \frac{1}{7} = \frac{3}{7}$$

*******   **Example**   *******   *******   ******

MULTIPLYING MIXED NUMBERS

The simplest way to multiply two mixed numbers is to change them to improper fractions and then multiply using cancellation wherever possible.

$$2\frac{1}{4} \times 5\frac{1}{3}$$

$$\frac{9}{4} \times \frac{16}{3} = \frac{3}{1} \times \frac{4}{1} = 12$$

1
To change $5\frac{2}{3}$ to an improper fraction, the denominator, 3, is multiplied by the whole number, 5, and the numerator, _____, is added.

2

$$5\frac{2}{3} = \frac{5 \times 3 + 2}{3} = \frac{15 + 2}{3} = \frac{17}{3}$$

2
Complete the following process of changing $6\frac{3}{5}$ to an improper fraction.

$$6\frac{3}{5} = \frac{6 \times 5 + 3}{5} = \frac{30 + 3}{5} = \underline{\hspace{1cm}}$$

$\frac{33}{5}$

3
Change $2\frac{1}{3}$ to an improper fraction. _____

$\frac{2 \times 3 + 1}{3} = \frac{7}{3}$

4
Change $4\frac{3}{8}$ to an improper fraction. _____

$\frac{4 \times 8 + 3}{8} = \frac{35}{8}$

5
Change $5\frac{3}{4}$ to an improper fraction. _____

$\frac{5 \times 4 + 3}{4} = \frac{23}{4}$

NEW IDEA

6
To multiply $2\frac{1}{3} \times 2\frac{1}{2}$ first change the mixed numbers to improper fractions.

$$2\frac{1}{3} = \frac{7}{3} \qquad 2\frac{1}{2} = \underline{\hspace{1cm}}$$

$\frac{5}{2}$

7
Complete the multiplication shown at the right. The mixed numbers have been changed to improper fractions.

$$2\frac{1}{3} \times 2\frac{1}{2}$$

$$\frac{7}{3} \times \frac{5}{2} = \underline{\hspace{1cm}}$$

$\frac{35}{6} = 5\frac{5}{6}$

8
To multiply $5\frac{1}{3} \times \frac{4}{5}$ first write $5\frac{1}{3}$ as an improper fraction.

$$5\frac{1}{3} = \underline{\hspace{1cm}}$$

$\frac{16}{3}$

142

9
Complete the multiplication shown at the right.

$5\frac{1}{3} \times \frac{4}{5}$

The mixed number has been changed to an improper fraction.

$\frac{16}{3} \times \frac{4}{5} = \underline{\hspace{1cm}}$

$\frac{64}{15} = 4\frac{4}{15}$

10
Multiply $6\frac{1}{2} \times \frac{3}{4}$ by first changing the mixed number to an improper fraction. $\underline{\hspace{1cm}}$

$\frac{13}{2} \times \frac{3}{4} = \frac{39}{8} = 4\frac{7}{8}$

11
Multiply $3\frac{1}{2} \times \frac{1}{2}$. $\underline{\hspace{1cm}}$

$\frac{7}{2} \times \frac{1}{2} = \frac{7}{4} = 1\frac{3}{4}$

12
$\frac{3}{5} \times 4\frac{1}{2} = \underline{\hspace{1cm}}$

$\frac{3}{5} \times \frac{9}{2} = \frac{27}{10} = 2\frac{7}{10}$

NEW IDEA

13
Cancellation is a valuable shortcut in multiplying mixed numbers. Use cancellation to complete the problem shown below.

$2\frac{1}{3} \times \frac{2}{7} = \frac{7}{3} \times \frac{2}{7} = \underline{\hspace{1cm}}$

$\frac{1}{3} \times \frac{2}{1} = \frac{2}{3}$

******* **Example** ******* ******* ******

DIVIDING MIXED NUMBERS

To divide two mixed numbers, first change the mixed numbers to improper fractions.

$3\frac{1}{5} \div 2\frac{1}{3}$

Then complete the problem by multiplying the dividend by the divisor's reciprocal.

$\frac{16}{5} \div \frac{7}{3}$

The example at the right shows the steps in dividing two mixed numbers.

$\frac{16}{5} \times \frac{3}{7} = \frac{48}{35} = 1\frac{13}{35}$

☀☀☀☀☀☀☀   **Example**   ☀☀☀☀☀☀☀   ☀☀☀☀☀☀☀   ☀☀☀☀☀☀

CHECKING DIVISION

The long division problem shown at the right
is checked by multiplying the divisor and
the quotient.

$$\begin{array}{r} 575 \\ 7\,\overline{)4025} \end{array}$$

$$\begin{array}{r} \text{Check} \\ 575 \\ \underline{\times\ 7} \\ 4025 \end{array}$$

The check shows the division to have been done correctly because the
product of the divisor and quotient is the dividend.

Division of fractions is checked in exactly the same way. When the
product of the divisor and quotient is the dividend, the division
checks.

ⁿⁿⁿⁿⁿⁿⁿⁿⁿ   ⁿⁿⁿⁿⁿⁿⁿⁿ   ⁿⁿⁿⁿⁿⁿⁿⁿⁿ   ⁿⁿⁿⁿⁿⁿⁿⁿ   ⁿⁿⁿⁿⁿⁿⁿ

14

$$\frac{5}{7} \div \frac{3}{4} = \frac{5}{7} \times \frac{4}{3} = \frac{20}{21}$$

Is the product of the divisor, $\frac{3}{4}$, and quotient, $\frac{20}{21}$,
equal to the dividend? _____

Yes, $\frac{3}{4} \times \frac{20}{21} = \frac{5}{7}$

15

Show the check for the problem below.

$$\frac{7}{9} \div \frac{2}{3} = \frac{7}{9} \times \frac{3}{2} = \frac{7}{6}$$ _____

$\frac{2}{3} \times \frac{7}{6} = \frac{7}{9}$

16

Divide and check. $\frac{8}{9} \div 6 = $ _____

$\frac{4}{27}$ and $6 \times \frac{4}{27} = \frac{8}{9}$

17

Divide and check. $\frac{3}{8} \div \frac{3}{4} = $ _____

$\frac{1}{2}$ and $\frac{3}{4} \times \frac{1}{2} = \frac{3}{8}$

18

Divide and check. $4 \div \frac{3}{5} = $ _____

$\frac{20}{3}$ and $\frac{3}{5} \times \frac{20}{3} = 4$

19
Divide and check. $\frac{4}{9} \div 2\frac{1}{3} =$ _____

$\frac{4}{21}$ and $\frac{7}{3} \times \frac{4}{21} = \frac{4}{9}$

20
Divide and check. $1\frac{1}{4} \div 2\frac{3}{5} =$ _____

$\frac{25}{52}$, $\frac{13}{5} \times \frac{25}{52} = \frac{5}{4}$

******* **Example** ******* ******* ******

A FRACTION "OF" A NUMBER MEANS MULTIPLICATION

The phrase **"a fraction of a number"** always can be interpreted as indicating multiplication of the fraction and the number.

$\frac{3}{4}$ **of** 28   means $\frac{3}{4} \times 28$.

$\frac{7}{5}$ **of** 46   means $\frac{7}{5} \times 46$.

21
To find $\frac{3}{5}$ **of** 6, use the fact that the phrase

"$\frac{3}{5}$ **of** 6"   means $\frac{3}{5} \times 6$

To find $\frac{2}{3}$ **of** 8, _____ (multiply, add) the numbers.

multiply

22
To find $\frac{7}{8}$ **of** 5, multiply $\frac{7}{8} \times 5$. What is $\frac{7}{8}$ **of** 5?

_____

$\frac{7}{8} \times \frac{5}{1} = \frac{35}{8} = 4\frac{3}{8}$

23
To find $\frac{9}{10}$ of 15, _____ (divide, multiply) $\frac{9}{10}$ and 15.

multiply

24

Find $\frac{9}{10}$ of 15. _____ $\frac{9}{10}$ x $\frac{15}{1}$ = $\frac{27}{2}$ = 13 $\frac{1}{2}$

25

Find $\frac{3}{4}$ of 8. _____ $\frac{3}{4}$ x $\frac{8}{1}$ = 6

26

Find $\frac{5}{8}$ of 24. _____ $\frac{5}{8}$ x $\frac{24}{1}$ = 15

27

Suppose $\frac{3}{4}$ of a class of 20 students were absent.

How many students were absent? _____ $\frac{3}{4}$ x 20 = 15

28

If a carpenter has finished $\frac{2}{3}$ of a 30 hour job,

how much has he finished? _____ $\frac{2}{3}$ x 30 = 20

29

If a student has read $\frac{5}{6}$ of his 120 page reading lesson,

how many pages have been read? _____ $\frac{5}{6}$ x 120 = 100

******   **Example**   ******   ******   ******

DIVISION IS INDICATED BY THE QUESTION "HOW MANY ARE IN?"

The problem:

      "How many 5 cent pencils can be purchased for 85 cents?"

can be re-phrased as

"How many 5's are in 85?"

The question is answered by dividing 85 by 5.

Whenever a problem asks:

      "How many of a first number are in a second number?"

the answer is found by dividing the first number into the second.

30

To find

How many $\frac{1}{2}$ inches are in 8 inches

8 is divided by $\frac{1}{2}$. Which of the following shows

8 divided by $\frac{1}{2}$? _____

   a) $\frac{1}{2} \div 8$ b) $8 \div \frac{1}{2}$ b

31

How many $\frac{1}{2}$ inches are in 8 inches? _____ $8 \div \frac{1}{2} = 16$

32

To find how many $\frac{1}{4}$ years there are in 5 years,

_____ (multiply, divide).   divide

33

Find how many $\frac{1}{4}$ years there are in 5 years. _____ $5 \div \frac{1}{4} = 20$

34

Find how many $\frac{1}{3}$ acres there are in 12 acres. _____ $12 \div \frac{1}{3} = 36$

35

Find how many $\frac{1}{2}$ ounces there are in 32 ounces. _____ $32 \div \frac{1}{2} = 64$

36

How many $\frac{1}{4}$ inches are in 3 inches? _____ $3 \div \frac{1}{4} = 12$

37

How many $\frac{1}{10}$ grams are in 12 grams? _____ $12 \div \frac{1}{10} = 120$

38

How many $\frac{4}{5}$ liters are in 7 liters? _____ $7 \div \frac{4}{5} = \frac{35}{4} = 8\frac{3}{4}$

39

The problem:

   How many $\frac{3}{5}$ gram rings can be made from 10

   grams of gold?

can be re-phrased as:

   How many _____ are in _____ ? $\frac{3}{5}$, 10

40

How many $\frac{3}{5}$ gram rings can be made from 10 grams of gold? _____

$$10 \div \frac{3}{5} = \frac{50}{3} = 16\frac{2}{3}$$

16 rings

41

The problem:

  Find the number of $\frac{7}{10}$ grams pills that can be made from 21 grams of medicine

can be rephrased to:

  How many _____ are in _____ ?

$\frac{7}{10}$, 21

42

Find the number of $\frac{7}{10}$ gram pills that can be made from 21 grams of medicine. _____

$$21 \div \frac{7}{10} = 30$$

43

How many $3\frac{4}{5}$ kilogram bricks can be made from 100 kilograms of concrete? _____

$$100 \div 3\frac{4}{5} = 26\frac{6}{19}$$

26 bricks

44

Razor blades weighing $1\frac{3}{10}$ grams each are to be made from 91 grams of steel. How many can be made? _____

$$91 \div 1\frac{3}{10} = 70$$

45

If capsules containing $3\frac{4}{5}$ milligrams of medicine are to be filled from a supply of 76 mg, how many can be made? _____

$$76 \div 3\frac{4}{5} = 20$$

46

How many home mortgages are there if the average interest per year is $\$1\frac{1}{2}$ thousand for each mortgage and the total interest collected per year is \$45 thousand? _____

$$45 \div 1\frac{1}{2} = 30$$

Post Quiz #2

This quiz reviews the preceding unit. Answers are at the back of the book.
Do not continue until all problems are understood.

Divide and check

1 $\frac{6}{7} \div \frac{3}{5}$

2 $2\frac{1}{4} \div 4\frac{1}{2}$

3 Find $\frac{4}{7}$ of 21.

4
Find how many $\frac{1}{3}$ years there are in 6 years.

5
Find how many $\frac{1}{2}$ quart bottles can be filled with 8 quarts of liquid.

6
A general estimates that $\frac{3}{8}$ of his 32 battalions are well prepared.
How many battalions are well prepared?

7
How many $\frac{1}{4}$ ounce coins can be made from 25 ounces of silver?

8
How many $\frac{3}{5}$ kilogram blocks of lead can be made from 63 kilograms of lead?

9
Bill has to put $\frac{3}{5}$ of a 7 ton load of sand on a vacant lot. How many
tons of sand should he put on the lot?

10
John spends $80 to buy stock that costs $$5\frac{3}{8}$ per share. How many shares
does he buy?

Chapter 8

UNIT 3: WORD PROBLEMS REQUIRING ADDITION
OF DECIMAL NUMERALS

PLACE VALUES OF DECIMAL FRACTIONS

The diagram below shows the names of the place values right of the decimal.

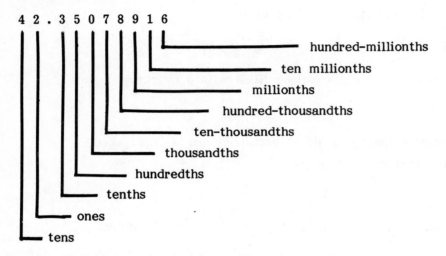

4 2 . 3 5 0 7 8 9 1 6

hundred-millionths

ten millionths

millionths

hundred-thousandths

ten-thousandths

thousandths

hundredths

tenths

ones

tens

There are several important lessons to be learned about the place value names.

1)   Each place is one-tenth as great as the place on the left.

$\frac{1}{10}$ is one-tenth as great as 1.

$\frac{1}{100}$ is one-tenth as great as $\frac{1}{10}$.

2)   There is no one's place on the right of the decimal.

1

The word **"sum"** means the answer to an addition problem.

George has 1.05 kilograms of aspirin and 0.98 kilograms of aspirin. Find the sum of the weight of the drug. _____

```
  1.05
+0.98
  2.03  kilograms
```

2

Find the **sum** of 5.6 milligrams (mg) and 23.92 mg._____

```
  5.60
+3.92
  9.52  mg
```

3

What is the **sum** of 0.45 liters and 1.9 liters? _____

```
  0.45
+1.90
  2.35  liters
```

4

1.4 kilograms and 0.5 feet cannot be added because the numbers have different labels.
Can the sum of 483.25 grams and 37.099 grams (gm) be found? _____

```
Yes,
  483.25
+  37.099
  520.349  gm
```

5

Can 57.3 pounds be added to 0.49 feet? _____

No, the labels are not the same.

6

Nurse Reynolds needed to give dosages of 6 minims, 28 minims, and 14 minims. Find the sum of the dosages. _____

```
   6
  28
+14
  48  minims
```

7

Cotes Sporting Goods store received two shipments costing $1,468.05 and $82,025.36. Find the sum of the costs of the shipments. _____

```
   1,468.05
+82,025.36
$83,493.41
```

8

Find the sum of forces on an object if they are 42.5 kilograms (kgm), 1.07 kgm, 0.78 kgm, and 38 kgm. _____

```
  42.50
   1.07
   0.78
+38.00
  82.35  kgm
```

9

A pharmacy received the following amounts of
alcohol: 4.05 liters of alcohol, 0.561 liters, and
0.26 liters. Find the sum of the amounts of
alcohol. _____

```
  4.050
  0.561
+0.260
  4.871 liters
```

10

Bill ran a mile race with the following quarter mile
split times: 58.3 sec, 61.1 sec., 58 sec., 59.28 sec.
Find the sum of the split times. _____

```
  58.03
  61.10
  58.00
+59.28
 236.41 sec
```

11

Find the sum of the commission fees for a job
when the four fees were 1.936%, 2.05%, 2.19%,
and 2.335%. _____

```
  1.936
  2.050
  2.190
+2.335
  8.511%
```

ⅢⅢⅢⅢⅢ   ⅢⅢⅢⅢⅢ   ⅢⅢⅢⅢⅢ   ⅢⅢⅢⅢⅢ   ⅢⅢⅢⅢⅢ

Post Quiz #3

This quiz reviews the preceding unit. Answers are at the back of the book.
Do not continue until all problems are understood.

1
Add 28.026 and 37.082

2
Add 1,350.5 and 2,259.21

3
An architect's fees were 3.975% and 4.1092%. Find the sum of the fees.

4
The ingredients in a medicine tablet weighed 2.03 mg, 14 mg, 8.785 mg
and 0.573 mg. Find the sum of the weights of the ingredients.

5
The amounts of three solutions to be mixed are 4.5 liters, 9.236
liters, and 2 liters. What is the sum of the three amounts?

6
The unit prices of 3 sizes of screws were found to be $.0397, $.125,
and $.38275. Find the sum of the prices.

UNIT 4: WORD PROBLEMS REQUIRING SUBTRACTION
OF DECIMAL NUMERALS

*******   **Example**   *******   *******   ******

CHECKING SUBTRACTION

In the subtraction example on the right:

0.50 is the **minuend**	0.50
0.12 is the **subtrahend**	−0.12
and 0.38 is the **difference.**	0.38

The subtraction is checked by the addition shown at the right. The sum of the difference and the subtrahend must be the minuend.

0.38	**difference**
+0.12	**subtrahend**
0.50	**minuend**

ᴨᴨᴨᴨᴨᴨᴨ   ᴨᴨᴨᴨᴨᴨᴨᴨ   ᴨᴨᴨᴨᴨᴨᴨ   ᴨᴨᴨᴨᴨᴨᴨ   ᴨᴨᴨᴨᴨ

1
In the subtraction example on the right, the **minuend** is 0.03, the **subtrahend** is 0.007 and the **difference** is 0.023. Add the **difference** and **subtrahend**. Is the sum equal to the **minuend**? _____

0.03
−0.007
0.023

Yes
0.023
+0.007
0.030

2
Does the subtraction problem on the right check? _____

Hint: Add the **difference** and the **subtrahend.**

1.0278
−0.11
0.9178

Yes
0.11
+0.9178
1.0278

3

In the subtraction 0.008 - 0.001 = 0.007,
      the minuend is _____,   0.008
      the subtrahend is _____,   0.001
 and the difference is _____.   0.007

4

In the subtraction, "take 0.02 from 0.08"
      the minuend is _____,   0.08
      the subtrahend is _____,   -0.02
      and the difference is _____.   0.06

5

In the subtraction "take 1.9 from 2.9"
      the minuend is _____,   2.9
      the subtrahend is _____,   -1.9
      and the difference is _____.   1.0

6

If the minuend is 10.4 and the subtrahend is 4.3,   10.4
what is the difference? _____   -4.3
                                                      6.1

NEW IDEA

7

    0.08   minuend
  -0.019   subtrahend
   0.061   difference

If the sum of the difference and the subtrahend is
the minuend a subtraction problem checks. Does the
problem above check? _____   Yes, 0.061
                                                +0.019
                                                0.080

8

   0.023   0.019
  -0.014   +0.014
   0.019   0.033

Does the subtraction problem above check? _____   No, 0.023 ≠ 0.033

9

$$10.9 - 1.8 = 9.1$$

In the problem above:

the minuend is	10.9
the subtrahend is	1.8
and the difference is	9.1

Does the difference check? _____

Yes, 9.1
+1.8
10.9

10

Subtract and check: $0.5 - 0.2 =$ _____

0.5   0.3
−0.2 +0.2
0.3   0.5

11

A subtraction problem checks when

_____ + _____ = _____

difference plus
subtrahend = minuend

NEW IDEA

12

The word **"difference"** is the name of the answer to a subtraction problem.

If one book weighs 28.3 kilograms and another book weighs 16.2 kilograms (kgm), what is the difference of the weights? _____

28.3
−16.2
12.1

NEW IDEA

13

The difference cannot be found between 2 dozen eggs and 1.9 grams of butter because the labels are not the same. Can the difference between 48.2 milliliters (ml) and 39.1 milliliters be found? _____

Yes, 48.2
−39.1
9.1 ml

Chapter 8

NEW IDEA

14
The phrase

"how much more"

implies a subtraction situation.

Jose has $1,268.09 in his savings account and Maria has $932.78 in hers. How much more is in Jose's account than in Maria's? _____

$$\begin{array}{r} 1,268.09 \\ -\ 932.78 \\ \hline 335.31 \end{array}$$

15
Sheerif has $92.56 and Abdullah has 48 kopeks. How much more money does Sheerif have than Abdullah? _____

Cannot be found
Labels are different

NEW IDEA

16
The phrase

"how much was left"

indicates a subtraction situation.

$15,600 was withdrawn from a bank account of $92,500. How much was left? _____

$$\begin{array}{r} 92,500 \\ -15,600 \\ \hline \$76,900 \end{array}$$

17
A supply of 38.5 liters of solvent was decreased by 10.6 liters. How much was left? _____

$$\begin{array}{r} 38.5 \\ -10.6 \\ \hline 27.9 \end{array} \text{ liters}$$

NEW IDEA

18
The word **"less"** implies subtraction.

4 less than 9 means 9 - 4

Does the sum of 8 and 2 mean 8 - 2 ? _____

No, 8 + 2

19
A first tablet weighed 3.9 milligrams (mgm) less than another tablet that weighed 12.1 milligrams. Find the weight of the first tablet. _____

$$\begin{array}{r} 12.1 \\ -3.9 \\ \hline 8.2 \end{array} \text{ mgm}$$

20

The interest rate on Iris's savings account is
1.42% per quarter. Percy's account pays 1.53%
per quarter. How much more is Percy's interest
rate than Iris'? _____

$$\begin{array}{r} 1.53 \ \% \\ -1.42 \ \% \\ \hline 0.11 \ \% \end{array}$$

21

If Vito's salary is raised to $6.85 per hour and
Luigi makes $5.87 per hour, how much more does
Vito make each hour than Luigi? _____

$$\begin{array}{r} 6.85 \\ -5.87 \\ \hline \$0.98 \end{array}$$

22

Sammy makes $1.25 less per hour than Johnny
who makes $8.19 per hour. How much does Sammy
make per hour? _____

$$\begin{array}{r} 8.19 \\ -1.25 \\ \hline \$6.94 \end{array}$$

23

A virus traveled 0.0068 millimeters (mm) through
a human cell which had a thickness of 0.0092
millimeters. How much farther does the virus
need to go to travel completely through the
cell? _____

$$\begin{array}{r} 0.0092 \\ -0.0068 \\ \hline 0.0024 \ mm \end{array}$$

24

A specimen of rock feldspar weighs 106.2
milligrams and a specimen of lignite weighs
28.3 milligrams. Find the difference in
their weights. _____

$$\begin{array}{r} 106.2 \\ -28.3 \\ \hline 77.9 \ mgm \end{array}$$

25

A supply of 42 cubic centimeters (cc) of morphine
was decreased by 9.7 cubic centimeters. How much
was left? _____

$$\begin{array}{r} 42.0 \\ -9.7 \\ \hline 32.3 \ cc \end{array}$$

Chapter 8

Post Quiz #4

This quiz reviews the preceding unit.   Answers are at the back of the book.
Do not continue until all problems are understood.

1
```
   1.27
 -0.16
```

2
```
  0.0056
 -0.0019
```

3
```
  0.5
 -0.13
```

4
```
 128.57
 -75.2
```

5
4.75 - 2.92 = _____

6
0.406 - 0.01932 = _____

7
0.0125 - 0.0118 = _____

8
Take 0.7 from 12.2

9
Find the difference between 4.0798 and 3.526.

10
If a subtraction problem has a difference of 12.05 and a subtrahend
of 10.39, what is the minuend?

11
Joe's specimen weighed 28.5 grams and Bill's weighed 37.2 grams.
How much more did Bill's weigh than Joe's?

12
A jar of 0.546 liters of alcohol was drained from a container
holding 2.729 liters. How much was left in the container?

UNIT 5: WORD PROBLEMS REQUIRING MULTIPLICATION
OF DECIMAL NUMERALS

******* Example ******* ******* ******

MULTIPLYING DECIMAL FRACTIONS

The decimal point is placed according to the total number of digits right of the decimal in the two factors of the problem.

The example at the right shows the multiplication of 8.5 by 0.68.

```
      8.5
  x 0.6 8
    6 8 0
    5 1 0
    0 0
  5.7 8 0
```

Notice that there are three decimal places in the product because the total number of digits right of the decimal points in the multiplier and the multiplicand is three.

8.5 x 0.68 = 5.780 or 5.78

******* Example ******* ******* ******

THE CLUE WORD "EACH" FOR MULTIPLICATION

In the problem, "What is the cost of 120 books if **each** book costs $4.95?" the word "**each**" gives the clue to multiply to find the answer.

120 x $4.95 = $594.00 total cost

1
The word "**each**" often indicates a multiplication situation. If 250 test tubes hold 102.5 milliliters (ml) **each,** the total number of milliliters that the test tubes can hold is the product of 250 and 102.5. How many milliliters can the tubes hold? _____   25,625 ml

2
If 4,000 razor blades weigh 0.129 grams **each,** what
is the total weight of the blades? _____

516.000 = 516 gm

3
If **each** theater ticket costs $3.50, how much do
2,000 cost? _____

$7,000.00

4
If the fee for **each** dollar value of a travelers
check is $0.0075, what is the fee for $300 worth of
travelers checks? _____

$2.2500 or $2.25

5
A bank charges $0.11 for each dollar loaned for one
year. What will it charge for loaning $12,000 for
one year? _____

$1,320.00

6
A nurse has to find the amount of aspirin in 300
tablets that contain 0.05 grams of aspirin each.
Find the weight of aspirin in the tablets. _____

15 gm

7
A test showed that each liter of air in a person's
lung contained 0.09 mg of pollutants. If Sue's
lungs hold 5 liters of air, find the amount of
pollutants in her lungs. _____

0.45 mg

******* Example ******* ******* ******

"OF" AS A CLUE WORD FOR MULTIPLICATION

To find "A fraction **of** a number" the fraction is always multiplied
by the number.

Decimal numerals are fractions and any sentence of the form

"A decimal numeral of a number"

indicates that the decimal numeral should be multiplied by the number.

8

To find "0.8 of 53" the word "**of**" indicates that
0.8 should be multiplied by 53.

Find 0.8 **of** 53. _____ 　　　　　　　　　42.4

9

Find 0.3 **of** 4.8 by multiplying the numbers. _____ 　　1.44

10

Find 0.079 of 11,000. _____ 　　　　　　　　869.000

11

Find 0.0037 of 4.6. _____ 　　　　　　　　　0.01702

12

In the problem "John lost .34 of his $300. How much
money did he lose?" the word "**of**" indicates that
0.34 and 300 should be _____. 　　　　　multiplied

13

A general estimates that .35 of his 2000 rifles are
defective. How many rifles are defective? _____ 　.35 x 2000 = 700

14

An engineer has .42 of his 40 day project completed.
How many days are completed? _____ 　　　.42 x 40 = 16.8

15

　　　　0.05 of a vat of beer is sediment. The vat
holds 5,000 liters. The "of" phrase "0.05 of a vat
of beer" is equivalent to the phrase

　　　　0.05 of 5,000 liters

because the vat holds _____ liters. 　　　　　5,000

16

　　　　1.2 of the cost of a refrigerator is the sale
price. The cost is $300. From the information in
the two previous sentences the "of" phrase

　　　　"1.2 of $300"

can be made. The words "the cost" in the first
sentence were replaced by _____. 　　　　　$300

17

0.6 of the people seeing a doctor have no physical problems. A doctor sees 60 people. The phrase "0.6 of the people" can be replaced by the "of" phrase:

0.6 of _____.

60 people

18

0.875 of the volume of milk is water. A tank truck holds 20,000 liters of milk. The phrase

"0.875 of the volume of milk" can be replaced by the "of" phrase _____.

0.875 of 20,000 liters

19

Ariane has 300 ml of a medicine. 0.3 of the medicine is a solvent. Using the information in the two sentences above, the phrase "0.3 of the medicine" can be replaced by the "of" phrase _____.

0.3 of 300 ml

20

0.095 of the price of a car goes toward the administrative expense of the company. The price of a car is $5,900. $5,900 can be substituted for "price" in the phrase
0.095 of the price
to obtain the "of" phrase _____.

0.095 0f $5,900

21

0.116 of the amount of a mortgage is the interest on the mortgage for one year. The mortgage is for $30,000. How much is the interest? _____

0.116 of $30,000
$3480.00

22

A microwave oven can cook a roast in 0.6 of the time needed by an electric oven. An electric oven cooks a roast in 3 hours. How much time does the microwave oven need to cook the roast? _____

0.6 of 3 hours
1.8 hours

23

0.21 of the atmosphere is oxygen. A closet contains 3 cubic meters of air. How much of the air is oxygen? _____

0.21 of 3 cu. m
0.63 cu. m
of oxygen

24
The selling price of a book is $20.50. The printing cost is 0.12 of the selling price. Find the printing cost of the book. _____

0.12 of $20.50
$2.46

25
A diver's tank will hold 203.8 liters of air. 0.79 of the volume of the air is nitrogen. How much of the air in the tank is nitrogen? _____

161.002 liters

26
0.8 of a bottle of medicine is codeine. The bottle contains 200 drams of medicine. How much codeine is in the bottle? _____

160 drams

27
A loan payment is 1.3 of the principal part of the payment. The principal part of the payment is $80. Find the size of the loan payment. _____

$104.00

Chapter 8

Post Quiz #5

This quiz reviews the preceding unit. Answers are at the back of the book.
Do not continue until all problems are understood.

1
If 25 radios cost $9.50 each, what is the cost of all the radios?

2
Each cubic meter of gas weighs 1.56 kgm. A tank holds 40 cubic meters
of the gas. Find the weight of the gas in the tank.

3
A department store charges $0.015 a month for each dollar it loans. Janie
borrows $300 to buy a set of silver. How much is the interest on the
loan for one month?

4
Sam buys 24 peaches. They cost $0.12 each. How much will he pay for
the peaches?

5
A processed meat product weighs 2.5 kgm. Each kgm of the product is
allowed to have 0.0009 kgm of insect parts. How many kgm of insect
parts could be in the product?

6
    0.3 of 8 = _____

7
    1.6 of 500 = _____

8
    0.2 of a tablet is caffeine. The tablet weighs 5 milligrams.
How much caffeine is in the tablet?

9
The price of brand X deodorant is 1.13 of the price of Snickle
deodorant. Snickle costs $2 per can. What is the price of brand X?

10
In a genetics experiment 200 brown rats were used. 0.87 of the
rats had brown eyes. How many rats had brown eyes?

11
Sam is to receive a raise of 0.15 of his salary for the previous year.
Find the size of the raise if his salary last year was $22,000.

164

UNIT 6: WORD PROBLEMS REQUIRING DIVISION
OF DECIMAL NUMERALS

******* Example ******* ******* ******

ROUNDING OFF DECIMAL DIVISION QUOTIENTS

Division answers (quotients) are often approximated by rounding off the answer at a particular number of decimal places. The example below illustrates the procedure for rounding off the answer of $1 \div 3$ at two decimal places:

1) Since the quotient is to be rounded off at **two** places, the division of 1 by 3 is carried to **three** decimal places.

2) The decimal 0.333 has the digit "3" in its third decimal place and is rounded off to 0.33.

```
        . 3 3 3
   3 )1.0 0 0
      - 9
        1 0
        - 9
          1 0
          - 9
            1
```

If the digit in the third place is less than 5 then it is dropped and the second digit remains unchanged.

$$0.333 \doteq 0.33$$

The symbol "$\doteq$" means "approximately equal to."

******* Example ******* ******* ******

SOLVING WORD PROBLEMS USING DIVISION OF DECIMALS

Word problems using the word **"each"** frequently are solved by division.

When numbers are given for the **total** and for **each** item that makes up the total, the word problem is solved by dividing the **"total"** number by the **"each"** number.

$$(\text{"total" number}) \div (\text{"each" number}) = \text{answer}$$

Chapter 8

1

Beer is sold by the case or by the can, but the cost is the same.   If **each** can costs \$.60 and a case (**total**) costs \$14.40 find the number of cans in a case. _____

14.40 ÷ .60
= 24 cans

2

Homer has a loan at the bank which costs him \$.13 interest **each** day.   When his **total** interest is \$4.55 how many days have passed? _____

4.55 ÷ .13
= 35 days

3

A pile of coins weighs 243.6 grams (gm).   If each coin weighs 1.2 gm, how many coins are in the pile? _____

243.6 ÷ 1.2
= 203 coins

4

A carnival sold tickets for \$.75 each.   At the end of the performance the box office had a total of \$51.00.   How many tickets had been sold? _____

51 ÷ .75
= 68 tickets

5

A manufacturing plant uses 4.2 cm of wire on each item.   If 3549 cm of wire are used in a day, how many items are manufactured? _____

3549 ÷ 4.2
= 845 items

6

A new carpet is purchased for \$9 a square yard. The total bill was \$725.40.   How many square yards were purchased? _____

725.40 ÷ 9
= 80.6 sq. yds.

******* **Example** *******   *******   ******

AVERAGES ARE FOUND BY USING DIVISION

To find the **average** of a group of numbers, first add the numbers and then divide the sum by the number of addends.

(total or sum) ÷ (number of addends) = average

If the division does not end with a zero remainder then round off the quotient at two decimal places.

7
Tom has taken four math tests and received scores
of 92, 85, 67, and 82. Find his **average** by
adding the numbers and dividing by 4. _____

$326 \div 4 = 81.5$

8
The seven linemen on the football team weigh 183
lbs, 202 lbs, 175 lbs, 210 lbs, 194 lbs, 183 lbs,
and 198 lbs. Find the **average** weight by
adding the weights and dividing by 7. _____

$1345 \div 7$
$= 192.14$ lbs.

9
Jake earned three paychecks over his Christmas
vacation of $75.63, $81.48, and $95.14. What was
his average paycheck? _____

$252.25 \div 3$
$= \$84.08$

10
Scott took a 5-day trip and registered his miles
traveled each day. His miles were 450, 292, 525,
139, and 387. What was his average miles traveled
each day? _____

$1793 \div 5$
$= 358.6$ miles

11
Six runners ran in a 3,000 meter race. Their
times, in minutes, were 10.3, 12.2, 19.6, 15.4,
14.7, and 11.8. What was the average time of
the runners? _____

$84.0 \div 6$
$= 14.0$ min.

12
The pollution in a stream was measured on five
consecutive days. The measures, in grams per liter,
were 14, 8, 25, 43, and 38. Find the average
pollution each day. _____

$128 \div 5 = 25.6$ gm
per liter

13
Sara mailed 45 packages at a total cost of $37.52.
What was the average mailing cost per package? _____

$37.52 \div 45 \doteq \$.83$

167

Chapter 8

14
Seven passengers boarded a plane. Their total
weight was 1083 pounds. What was the average
weight per passenger? _____

$$1083 \div 7$$
$$\doteq 154.71 \text{ lbs.}$$

15
Twelve packages weighing a total of 369 lbs
were loaded on a plane. What was the average
weight per package? _____

$$369 \div 12$$
$$= 30.75 \text{ lbs}$$

16
Sally earned $78, $93, $150, $225, and $392
on successive days as a waitress. What was her
average earnings per day? _____

$$\$938 \div 5$$
$$= \$187.60$$

Post Quiz #6

This quiz reviews the preceding unit. Answers are at the back of the book.
Do not continue until all problems are understood.

1
A small theater sold tickets for $.63 each, tax included, and collected
a total of $569.52. Find the number of tickets sold.

2
The heights of the members of the basketball team were, in meters,
1.84, 2.10, 1.92, 1.89, and 1.95. Find the average height of a player.

3
The druggist had sales of $243.85 last Monday from 95 separate purchases.
What was the average purchase price?

4
Smith bought 6.5 square yards of carpet for $54.60. What was the price
for each square yard of carpet?

CHAPTER 8 POST-TEST

This test reviews the objectives of the chapter. The student is expected to know how to do **all** of these problems before finishing the chapter. Answers for this test are at the end of the book.

1

Two rocks, one weighing $2\frac{3}{4}$ kilograms and the other weighing $1\frac{1}{2}$ kilograms, were placed on a scale. What was their total weight?

2

A piece of rope $5\frac{3}{4}$ meters long was cut from a rope $10\frac{1}{2}$ meters long. How much was left?

3

How much more is $5\frac{1}{8}$ cups than $3\frac{3}{8}$ cups?

4
In $\frac{3}{4} \div \frac{9}{8} = \frac{2}{3}$

    the divisor is _____
    the quotient is _____
    the dividend is _____

5
Find $\frac{7}{8}$ of 3.

6
A bank has $2\frac{7}{10}$ million dollars. $\frac{2}{3}$ of it was on deposit. How much was on deposit?

7
Spoons that weigh $\frac{1}{10}$ kilogram are to be made from $6\frac{3}{5}$ kilograms of silver alloy. How many spoons can be made?

8
The dealer cost of a car is $\frac{4}{5}$ of the selling price. The sales price is $3500. Find the dealer cost of the car.

9

In the problem $\frac{2}{3} \div \frac{5}{9} = \frac{6}{5}$,

the quotient is _____
the dividend is _____
the divisor is _____

10

Find $\frac{4}{5}$ of 25.

11

Bill had $3\frac{2}{3}$ liters of medicine. One-half of it was needed on a wound. How much was needed?

12

Clay cups weighing $\frac{3}{10}$ kilograms are to be made from $5\frac{7}{10}$ kilograms of clay. How many cups can be made?

13

The cost of a motorcycle is $\frac{5}{6}$ of the sales price. The sales price is $1800. Find the cost of the motorcycle.

14

In the statement 4 - 3 = 1
The difference is _____
the minuend is _____
the subtrahend is _____

15

The subtraction 361 - 294 = 67 checks when

_____ + _____ = _____

16

Take 2.58 from 9.26.

17

Find the difference between 1.903 and 1.295

18

What number is 1.09 less than 8.27?

19

If one capsule weighs 45.2 gm and another weighs 37.8 gm, how much more does the heavier capsule weigh than the lighter capsule?

20

Bill spends $10.28, $5.09, $105.56, and $28.97 on a shopping trip. Find the sum of his purchases.

21

One bank charges 0.0095 less interest per month than another bank which charges 0.0153 per month. Find the interest charge of the first bank.

22
If the letters A, B, and C represent three different numbers and A - B = C,
the minuend is _____,
the subtrahend is _____,
and the difference is _____.

23
The subtraction problem
378 - 188 = 190
checks when

_____ + _____ = _____.

24
If Sandra has $2,049.00 and Tom has $3,150.75, how much more does Tom have than Sandra?

25
The chemicals in a mixture weigh 0.25 kg, 1.6 kg, 0.097 kg, and 4.97 kg. Find the sum of the weights.

26
A small flask holds 4.51 liters less than a jar holding 8.49 liters. How much does the small flask hold?

27
Robert decreases his interest rate of 8.23% by 1.51%. What is his new interest rate after the decrease?

28
Find 0.37 of 56.

29
If 35 people are to receive dosages of nembutal of 7.48 mg each, how much of the medicine is needed?

30
The First Bank charges $0.109 per year for each dollar it loans. Robert and Linda need a $50,000 loan. How much will the bank charge to loan $50,000 for one year?

31
A force was directed so that 0.37 of the force moved an object along a plane. The force was 60 grams. How much of the force moved the object?

32
Stamps used to cost $.15 each. How many stamps could be bought with $6.45?

33
Sally keeps track of her grocery purchases for four consecutive weeks. She spends $87.52, $102.05, $48.26, and $78.09 in those four weeks. What was her average weekly grocery cost during this period?

34
Farmer Hank discovered that 0.43 of his grain had spoiled. Of his 60,000 bushels of wheat, how much had spoiled?

35
A pharmacist needs 25 doses of phenobarbital. Each dose is to be 5.72 milligrams. How much of this medicine is needed to fill the order?

36
The National Bank charges $0.115 per year for each dollar it loans. Burt and Lonny need a $30,000 loan. How much will the bank charge to borrow $30,000 for one year?

37
The fuel tank of a car holds 12 gallons. If the manufacturer uses thinner oil in the gearbox, the car will save 0.04 of its fuel. How much fuel from a full tank will be saved by using the lighter oil?

38
Christmas cards are on sale for $.18 each. How many can Pete buy with $16.92?

39
Max received four orders for $45.92, $87.50, $19.95, and $32.10. What was the average of these four orders?

Chapter 9

Anxiety Isn't an Addiction, But ...

I have a good friend who gave up smoking a few years ago, and received lots of support and encouragement for that decision. But two years later she "tried" one cigarette at a party and has been hooked ever since.

Math anxiety isn't an addiction, but once you start reducing it you must never allow yourself to get hooked again.

ᴨᴨᴨᴨᴨᴨ   ᴨᴨᴨᴨᴨᴨ   ᴨᴨᴨᴨᴨᴨ   ᴨᴨᴨᴨᴨᴨ   ᴨᴨᴨᴨᴨ

MAINTAINING CONTROL

You can maintain and strengthen your control over mathematics learning situations by:

1) Always using the Relaxation Response at the slightest feeling of anxiety.

2) Consciously noting the three steps in the learning process when confronted with any mathematics task.

3) Constantly practicing the four study skills of mathematics until they become habits.

UNIT 1: THE MEANING OF PERCENT

*******   **Example**   *******   *******   ******

PERCENT MEANS ONE-HUNDREDTHS

The idea of percent is that a quantity is divided into 100 equal parts. Each part represents one percent (1%) of the original quantity. The original quantity is 100%.

100 books are split into two piles; one pile with 83 books and the other with 17. The first pile then contains 83 percent of the 100 books and the second pile contains 17 percent of the original 100 books.

$200 was shared by two students, Tom and Mary. Tom received his share of $58 and Mary received $142. This means that if the original $200 were split into 100 equal parts then Tom had 29 of those parts ($58) or 29%. Similarly, Mary had 71 of those parts ($142) or 71%.

ᴨᴨᴨᴨᴨᴨ   ᴨᴨᴨᴨᴨᴨ   ᴨᴨᴨᴨᴨᴨᴨ   ᴨᴨᴨᴨᴨᴨ   ᴨᴨᴨᴨᴨ

1
The fraction $\frac{1}{5}$ means $1 \div 5$ or one whole divided into five equal parts. The fraction $\frac{1}{100}$ means $1 \div 100$ or one whole divided into _____ parts.   100

2
   $1 \div 100$ or $\frac{1}{100}$ is one-hundredth. One whole has been divided into _____ parts.   100

3
$\frac{41}{100}$ is read as 41 one-hundredths.

$\frac{1}{100}$ is read as 1 _____ - _____ .   one-hundredth

4
$\frac{7}{100}$ is read as 7 one-hundredths.

$\frac{11}{100}$ is read as 11 _____ - _____ .   one-hundredths

5
$\frac{23}{100}$ is read as 23 one-hundredths.

$\frac{47}{100}$ is read as 47 _____ - _____ .   one-hundredths

6
Percent means one-hundredths. 5 percent is the
same as 5 one-hundredths. 8 percent is the same
as 8 _____ - _____ .   one-hundredths

7
Percent means one-hundredths. 63 percent means
$\frac{63}{100}$ or 63 _____ - _____ .   one-hundredths

8
Percent means one-hundredths. 91 percent means
$\frac{91}{100}$ or 91 _____ - _____ .   one-hundredths

9
Percent means _____ - _____ .   one-hundredths

10
    47 one-hundredths is 47 percent.
53 one-hundredths is 53 _____ .   percent

11
The terms "percent" and "one-hundredths" are
interchangeable. 69 one-hundredths is the same
as 69 _____ .   percent

12
    59 one-hundredths and 59 percent have the
_____ (same, different) meaning.   same

177

13

      71 percent is the same as 71 _____ - _____ .   one-hundredths

14

      68 one-hundredths is the same as 68 _____ .   percent

15

If a quantity is divided into 100 equal parts each
part is equal to 1 percent. 7 of the parts would
be 7 _____ .   percent

16

If a quantity is divided into 100 equal parts then
37 of the parts is 37 _____ .   percent

17

If a quantity is divided into 100 equal parts, how
many of those parts represent 23 percent? _____   23

18

The symbol for percent is "%". 67 percent is
written as 67%. Use the percent symbol to write
47 percent. _____   47%

19

% is the symbol for _____ .   percent

20

Use the symbol % to write 96 percent. _____   96%

21

Use the symbol % to write 9 percent. _____   9%

22

The symbol for percent is _____ .   %

23

Write 56 percent in symbols. _____   56%

24
Write .3 percent in symbols. _____   .3%

25
Write 133 percent in symbols. _____   133%

26
Percent means hundredths.   51% means $\frac{51}{100}$.
19% is _____ . $\frac{19}{100}$

27
$\quad$ 41% is $\frac{41}{100}$. $\qquad$ 13% is _____ . $\frac{13}{100}$

28
$\quad$ 2% = $\frac{2}{100}$. $\qquad$ 1% = _____ . $\frac{1}{100}$

29
$\quad$ 9% = _____ . $\frac{9}{100}$

30
$\quad$ 7% = _____ . $\frac{7}{100}$

31
$\quad$ $\frac{2}{100}$ = 2%. $\qquad$ $\frac{18}{100}$ = _____ .   18%

32
$\quad$ $\frac{92}{100}$ = _____ .   92%

33
Write $\frac{150}{100}$ as a percent. _____   150%

34
Write $\frac{4}{100}$ as a percent. _____   4%

35
Write $\frac{1}{100}$ as a percent. _____   1%

179

Post Quiz #1

This quiz reviews the preceding unit. Answers are at the back of the book.
Do not continue until all problems are understood.

1
Write each of the following as a fraction.

a.   b.   c.   d.   e.
17%   49%   57%   143%   7%

2
Percent means _____ - _____ .

3
Write each of the following as percents.

a. $\dfrac{19}{100}$ b. $\dfrac{81}{100}$ c. $\dfrac{3}{100}$ d. $\dfrac{181}{100}$ e. $\dfrac{67}{100}$

UNIT 2: CHANGING PERCENTS TO DECIMAL FRACTIONS

******* Example ******* ******* ******

WRITING PERCENTS AS DECIMALS

100 cents is equal to $1.00   100% is equal to one whole (1.00).

In this section percents are written as decimals. The process used
is identical to the one used to change money given in cents to
money given in dollars.

1
11 cents = $.11   55 cents = $_____ $.55

2
To change cents to dollars the decimal point is
moved two places to the left. Write 85 cents or
85. cents as dollars. _____ $.85

3
        100 cents = $1.00   25 cents = $_____ $0.25

4
To change 52% to a decimal the decimal point is
moved two places to the left.
        52% = 52.% = .52   58% = 58.% = _____   .58

5
To change 9% to a decimal move the decimal point
two places to the left.
        9% = 9.% = .09   7% = 7.% = _____   .07

6
To change a percent to a decimal move the decimal
point two places to the left.
        92% = 92.% = .92   49% = 49.% = _____   .49

7
To change 5% to a decimal the decimal point is
moved two spaces to the left.
                5% = 5.% = _____   .05

8
Change 2% to a decimal by moving the decimal
point two places to the left. _____   2.% = .02

9
To change any percent to a decimal move the decimal
point _____ (how many) places left.   2

10
To change a percent to a decimal move the decimal
point two places _____ (right, left).   left

11
Write 14% as a decimal. _____   .14

12
Write 95% as a decimal. _____   .95

13
Write 57% as a decimal. _____   .57

14
Write 64% as a decimal. _____   .64

15
Write 8% as a decimal. _____   .08

16
Write 3% as a decimal. _____   .03

17
Write 140% as a decimal. _____   1.40

18
Write 107% as a decimal. _____   1.07

19
Write 2.6% as a decimal. _____   .026

20
Write 57.9% as a decimal. _____   .579

21
Write 78% as a decimal. _____   .78

22
Write 10% as a decimal. _____   .10

23
Write 248% as a decimal. _____   2.48

24
Write .6% as a decimal. _____   .006

25
Write 2% as a decimal. _____   .02

26
Write 110% as a decimal. _____   1.10

27
Write .1% as a decimal. _____   .001

28
Write 1.6% as a decimal. _____   .016

29
Write .03% as a decimal. _____   .0003

30
Write 1% as a decimal. _____   .01

ıııııııııı   ıııııııııı   ıııııııııı   ıııııııııı   ııııııııı

Post Quiz #2

This quiz reviews the preceding unit. Answers are at the back of the book.
Do not continue until all problems are understood.

Write each of the following as a decimal.

1	2	3	4	5
27%	85%	56%	4%	37%

6	7	8	9	10
163%	92%	5.8%	16%	47.3%

UNIT 3: CHANGING DECIMAL FRACTIONS TO PERCENTS

*******   **Example**   *******   *******   ******

WRITING DECIMALS AS PERCENTS

One dollar is equal to 100 cents.   One whole is equal to 100%.

In this section decimals are written as percents. The process used is identical to the process used to change money given as dollars to money given as cents.

1

    $1 is 100 cents $.63 is 63 cents

How many cents are in $.73? _____   73 cents

2

There are 100 cents in $1 and $1 = $1.00. Two decimal places are used to show cents.

$.41 is _____ cents.   41

3

To change a decimal to a percent move the decimal point two places to the right.

    .63 = 63%   .45 = _____%   45.% = 45%

4

To change a decimal to a percent move the decimal point two places to the right.

    .93 = 93%   .57 = _____%   57.% = 57%

5

Write .18 as a percent by moving its decimal point two places right. _____   18.% = 18%

6

Write .41 as a percent by moving its decimal point two places right. _____   41.% = 41%

7
Write .06 as a percent by moving its decimal point
two places right. _____   06.% = 6%

8
Write .09 as a percent. _____   09.% = 9%

9
Write .71 as a percent. _____   71%

10
Write 1.56 as a percent. _____   156%

11
Write .5 as a percent. _____   50.% = 50%

12
Write 6.2 as a percent. _____   620.% = 620%

13
Write .01 as a percent. _____   1%

14
    .002 = .2%
Write .007 as a percent. _____   .7%

15
Write 1.01 as a percent. _____   101%

16
Write .08 as a percent. _____   8%

17
Write .001 as a percent. _____   .1%

18
When converting a decimal numeral to a percent,
the point is moved _____ places to the _____ .   2, right

Chapter 9

19
Write .009 as a percent. _____   .9%

20
Write 1.62 as a percent. _____   162%

21
Write .0075 as a percent. _____   .75%

22
Write 1.25 as a percent. _____   125%

23
Write .0241 as a percent. _____   2.41%

24
Write .12 as a percent. _____   12%

      ⌐⌐⌐⌐⌐⌐⌐⌐   ⌐⌐⌐⌐⌐⌐⌐⌐   ⌐⌐⌐⌐⌐⌐⌐⌐   ⌐⌐⌐⌐⌐⌐⌐⌐   ⌐⌐⌐⌐⌐⌐⌐

Post Quiz #3

This quiz reviews the preceding unit.   Answers are at the back of the book.
Do not continue until all problems are understood.

Write each of the following as a percent.

1	2	3	4
.26	.89	.08	.63

5	6	7	8
2.4	.52	1.45	.22

9	10	11	12
.043	1.41	.01	.025

UNIT 4: CHANGING FRACTIONS TO PERCENTS

FRACTIONS AS PERCENTS

To change a fraction to its approximate percent:

1) Use long division to convert the fraction to a decimal correct to three decimal places.

2) Round off the decimal to two places.

3) Write the decimal as a percent.

The steps used to change the fraction $\frac{5}{7}$ to a percent are shown at the right.

$$\frac{5}{7} \doteq 71\%$$

```
  0.7 1 4 ≐ .71 = 71%
7 )5.0 0 0
  -4 9
     1 0
    - 7
     3 0
```

The symbol $\doteq$ means "approximately equals."

1
To change $\frac{2}{7}$ to its approximate percent the first step is to change $\frac{2}{7}$ to its decimal, correct to three decimal places. Write $\frac{2}{7}$ as a decimal. _____

$$\frac{0.285}{7\,)2.000} \doteq \frac{2}{7}$$

2
Round off .285 to two decimal places. Use the rule that if the third digit is 5 or more, then one (1) is added to the second digit. _____

.29

3
$$\frac{2}{7} \doteq .285 \doteq .29$$
Write .29 as a percent. _____

29%

4
Complete the first step in changing $\frac{3}{11}$ to its approximate percent by dividing $3 \div 11$ correct to three decimal places. _____

$$\frac{0.272}{11\,)3.000}$$

187

5

$$\frac{3}{11} \doteq .272$$

Round .272 to two decimal places. Use the rule that if the third digit is 4 or less, then it is dropped. _____

.27

6

$$\frac{3}{11} \doteq 0.272 \doteq 0.27$$

Write 0.27 as a percent. _____

27%

7

Complete the first step in changing $\frac{4}{7}$ to a percent by dividing $4 \div 7$ to three decimal places.

$$7 \overline{)4.000}^{\;0.571}$$

8

$$\frac{4}{7} \doteq .571$$

Round off .571 to two decimal places. _____

.57

9

$$\frac{4}{7} \doteq .571 \doteq 57$$

Write .57 as a percent. _____

57%

10

Use long division and rounding off to write $\frac{5}{11}$ as a percent. _____

$$11 \overline{)5.000}^{\;0.454 \;\doteq\; .45} \quad \text{or } 45\%$$

11

Use long division and rounding off to write $\frac{2}{3}$ as a percent. _____

$$3 \overline{)2.000}^{\;0.666 \;\doteq\; 67\%}$$

12

Use long division and rounding off to write $\frac{5}{8}$ as a percent. _____

$$8 \overline{)5.000}^{\;0.625 \;\doteq\; 63\%}$$

13
Use long division and rounding off to write $\frac{5}{6}$ as a percent. _____

$$6\overline{)5.000} \quad \frac{0.833}{} \doteq 83\%$$

14
Use long division and rounding off to write $\frac{1}{2}$ as a percent. _____

$$2\overline{)1.0} \quad \frac{0.5}{} = 50\%$$

15
Use long division and rounding off to write $\frac{7}{10}$ as a percent. _____

$$10\overline{)7.0} \quad \frac{0.7}{} = 70\%$$

******* **Example** ******* ******* ******

COMMON FRACTION — PERCENT EQUIVALENTS

Any fraction can be changed to a percent approximation using long division. However, some fractions occur frequently enough that their equivalents are often memorized.

These fractions are:

$\frac{1}{2} = 50\%$ $\qquad$ $\frac{1}{3} = 33\frac{1}{3}\%$ $\qquad$ $\frac{1}{4} = 25\%$ $\qquad$ $\frac{1}{5} = 20\%$

$\frac{1}{6} = 16\frac{2}{3}\%$ $\qquad$ $\frac{1}{7} = 14\frac{2}{7}\%$ $\qquad$ $\frac{1}{8} = 12\frac{1}{2}\%$ $\qquad$ $\frac{1}{10} = 10\%$

16
Write $\frac{9}{10}$ as a percent. _____

90%

17
Write $\frac{3}{8}$ as a percent. _____

38%

18
Write $\frac{3}{4}$ as a percent. _____

75%

19
Write $\frac{57}{80}$ as a percent using long division and rounding off. _____

$$80\overline{)57.000} \quad \frac{0.712}{} \doteq 71\%$$

20
Write $\frac{15}{47}$ as a percent using long division and rounding off. _____

$$47\overline{)15.000} \quad \frac{0.319}{} \doteq 32\%$$

21
Write $\frac{7}{104}$ as a percent. _____

$$104\overline{)7.000} \quad \frac{0.067}{} \doteq 7\%$$

22
$\frac{3}{5}$, $\frac{7}{11}$, and $\frac{85}{132}$

Every proper fraction is _____ (less, more) than one.

less

23
$$1 = 100\%$$
Every proper fraction is always less than 1 or less than _____%.

100

24
$\frac{57}{12}$ is an improper fraction.

Is $\frac{57}{12}$ less than 1? _____

no

25
$$1 = 100\%$$
$\frac{57}{12}$ is more than 1 or greater than _____%.

100

26
$\frac{9}{5}$ is _____ (more, less) than 100%.

more

27

Change $\frac{9}{5}$ to a percent using long division. _____

$$5\overline{)9.0} \quad \frac{1.8}{} = 180\%$$

28

Change $\frac{15}{8}$ to a percent using long division. _____

$$8\overline{)15.000} \quad \frac{1.875}{} \doteq 188\%$$

29

Change $\frac{23}{9}$ to a percent. _____

$$9\overline{)23.000} \quad \frac{2.555}{} \doteq 256\%$$

30

Change $\frac{63}{58}$ to a percent. _____

$$58\overline{)63.000} \quad \frac{1.086}{} \doteq 109\%$$

ᶥᶥᶥᶥᶥᶥᶥᶥ   ᶥᶥᶥᶥᶥᶥᶥᶥ   ᶥᶥᶥᶥᶥᶥᶥᶥ   ᶥᶥᶥᶥᶥᶥᶥᶥ   ᶥᶥᶥᶥᶥᶥ

Post Quiz #4

This quiz reviews the preceding unit.   Answers are at the back of the book. Do not continue until all problems are understood.

Write each of the following as a percent.

1	2	3	4	5
$\frac{3}{8}$	$\frac{5}{9}$	$\frac{4}{5}$	$\frac{1}{3}$	$\frac{3}{10}$

6	7	8	9	10
$\frac{15}{29}$	$\frac{41}{85}$	$\frac{19}{43}$	$\frac{57}{48}$	$\frac{6}{145}$

UNIT 5: THE THREE NUMBERS IN A PERCENT PROBLEM

*******   **Example**   *******   *******   ******

BASE, RATE, AND PART

Every percent problem has three numbers. They are the **base** (b), the **rate** (r), and the **part** (p).

In this unit, methods are explained for identifying these numbers in a percent problem statement.

ⲙⲙⲙⲙⲙⲙⲙⲙ   ⲙⲙⲙⲙⲙⲙⲙⲙ   ⲙⲙⲙⲙⲙⲙⲙⲙⲙ   ⲙⲙⲙⲙⲙⲙⲙⲙⲙ   ⲙⲙⲙⲙⲙⲙⲙ

1
There are three numbers in: 15% of 200 is 30.
The numbers are 15%, 200, and _____.

     30

2
How many numbers are in the following? _____
     12% of 400 is 48.

     3
     12%, 400, 48

3
How many numbers are in the following? _____
     60 is 5% of 1200.

     3
     60, 5%, 1200

NEW IDEA

4
A percent problem always involves three numbers.
One number is the **base** (b). The letter b stands
for the _____ of the problem.

     base

5
One number in a percent problem is the **base** (b).
The **base** is often indicated by the letter _____ .

     b

6
There are _____ numbers in a percent problem.

     three

NEW IDEA

7
The base is one number in a percent problem. The
rate (r) is another number. The letter r stands
for the _____ in a percent problem.   rate

8
One number in a percent problem is its base (b)
and another number is its **rate** (r). The **rate**
is often shown by the letter _____ .   r

9
There are _____ numbers in a percent problem.   three

NEW IDEA

10
In a percent problem one number is the base (b),
another is the rate (r), and the third number is
the **part** (p). The letter p stands for the
_____ in a percent problem.   part

11
       b is the base
       r is the rate
       p is the part

The part is shown by the letter _____ .   p

12
The three numbers in a percent problem are the
_____ (b), the rate (r), and the part (p).   base

13
The three numbers in a percent problem are the
base (b), the _____ (r), and the part (p).   rate

14
The three numbers in a percent problem are the
base (b), the rate (r), and the _____ (p).   part

15
The three numbers in a percent problem are the
_____, the _____, and the _____.   base, rate,
                                                        part

16
What number appears directly after the word
"**of**" in this percent statement? _____
      90% of 500 is 450.

500

17
What number appears directly after the word
"**of**" in this percent statement? _____
      24 is 40% of 60.

60

NEW IDEA

18
The base (b) number in a percent problem usually
follows the word "**of**." To find the base number
in a percent problem, look for the word "_____."

of

19
The word "of" in a percent problem is a good clue
for identifying the base (b). To find the base,
look for the word "_____."

of

20
Find the base (b) of the following percent statement
by choosing the number directly after the word "of."
      8% of 900 is 72. _____

b = 900

21
Find the base (b) of the following percent statement
by choosing the number directly after the word "of."
      25% of 44 is 11. _____

b = 44

22
Find the base of the following percent statement.
      30 is 75% of 40. _____

b = 40

23
Find the base of the following percent statement.
      125% of 40 is 50. _____

b = 40

24
Find the base (b) for:
      3 is 6% of 50. _____

b = 50

25

Find the base (b) for:

19% of 200 is 38. _____

b = 200

26

In the percent statement below the base is 35 and
the number with the percent symbol (%) is _____.

21 is 60% of 35.

60%

27

What number is with the percent symbol (%) in the
following statement? _____

25% of 60 is 15.

25%

NEW IDEA

28

The rate (r) in a percent statement is usually
indicated by the symbol %. To find the rate
number, look for the symbol _____ .

%

29

Identify the rate number in the following percent
statement by finding the number with the % symbol.

30% of 70 is 21.   _____

30%

30

Find the rate number in the statement.

16 is 50% of 32.   _____

50%

31

Find the rate number in:

18% of 300 is 54. _____

18%

32

Find the rate number in:

45 is 150% of 30. _____

150%

33

In a percent statement, the number that comes
directly after "of" is the base (b) and the number
with the percent symbol (%) is the _____ .

rate

34
Identify the base (b) and rate (r) in:
                    18 is 9% of 200. _____ _____
                                                        b = 200
                                                        r = 9%

35
Identify the base (b) and rate (r) in:
                    40% of 60 is 24. _____ _____
                                                        b = 60
                                                        r = 40%

36
The three numbers in a percent problem are the base
(b), the rate (r), and the part (p). Find the base:
        60% of 120 is 72.   _____
                                                        120

37
For the problem: 60% of 120 is 72.
        the base (b) = 120
        the rate (r) = _____
                                                        60%

38
For the problem: 60% of 120 is 72.
        the base (b) = 120
        the rate (r) = 60%
        the part (p) = _____
                                                        72

39
For the problem: 45 is 150% of 30.
        the base (b) = _____
                                                        30

40
For the problem: 45 is 150% of 30.
        the base (b) = 30
        the rate (r) = _____
                                                        150%

41
For the problem: 45 is 150% of 30.
        the base (b) = 30
        the rate (r) = 150%
        the part (p) = _____
                                                        45

42
Find the base, rate, and part for:

       55% of 80 is 44.

    b = _____   r = _____   p = _____   80, 55%, 44

43
Find b, r, and p for:

       85 is 17% of 500.

    b = _____   r = _____   p = _____   500, 17%, 85

44
Find b, r, and p for:

       150% of 90 is 135.

    b = _____   r = _____   p = _____   90, 150%, 135

45
Find b, r, and p for:

       67% of 200 is 134.

    b = _____   r = _____   p = _____   200, 67%, 134

46
Find b, r, and p for:

       4 is 200% of 2.

    b = _____   r = _____   p = _____   2, 200%, 4

47
Find b, r, and p for:

       3% of 8 is 0.24.

    b = _____   r = _____   p = _____   8, 3%, 0.24

48
Find b, r, and p for:

       14 is 20% of 70.

    b = _____   r = _____   p = _____   70, 20%, 14

Chapter 9

Post Quiz #5

This quiz reviews the preceding unit. Answers are at the back of the book.
Do not continue until all problems are understood.

1
The three numbers in a percent statement are called _____ (b),

_____ (r), and _____ (p).

2
To find the b in a percent statement, look for the word "_____".

3
To find the r in a percent statement, look for the symbol _____ .

4
Find b, r, and p for:

  a)
      53% of 500 is 265.

  b)
      17 is 85% of 20.

  c)
      125% of 60 is 75.

  d)
      144 is 80% of 180.

UNIT 6: FINDING THE PART WHEN THE BASE AND RATE ARE KNOWN

#######   Example   #######   #######   ######

USING BASE AND RATE TO FIND THE PART

When the base and rate are known, the part is found by multiplication.

(base) x (rate) = (part)   **b x r = p**

Before the multiplication is performed, the rate number must be written as a decimal or a fraction.

If b = 95 and r = 68%,

    1)   Write 68% as .68

    2)   Multiply 95 and .68

      p = 95 x .68 = 64.60

1
If b = 40 and r = 15%, the part is found using multiplication. Write 15% as a decimal by moving the decimal point two places to the left. _____

.15

2
If b = 40 and r = 15%,

    p = 40 x 15% = 40 x .15 = _____

6.00

3
When b = 40 and r = 15% then p = 6 because

    40 x 15% = 40 x .15 = 6.00 = 6

When b and r are known numbers then p is found by _____ (multiplication, division).

multiplication

4
To find p when b = 84 and r = 25%, first write 25% as a decimal or a fraction.

    25% = _____

.25 or $\frac{1}{4}$

5
If b = 84 and r = 25% then
        p = 84 x 25% = 84 x .25 = _____   21

6
If b = 60 and r = 25% then
        p = 60 x 25% = 60 x .25 = _____   15

7
Find p when b = 60 and r = 45%. _____   60 x .45 = 27

8
If b = 42 and r = 19% then
        p = 42 x 19% = _____ = _____   42 x .19 = 7.98

9
Find p when b = 89 and r = 68%. _____   89 x .68 = 60.52

10
Find p when b = 810 and r = 30%. _____   810 x .30 = 243

11
Find p when b = 746 and r = 7%. _____   746 x .07 = 52.22

12
Find p when b = 237 and r = 56%. _____   237 x .56 = 132.72

13
Find p when b = 612 and r = 3%. _____   612 x .03 = 18.36

14
Find p when b = 763 and r = 8%. _____   763 x .08 = 61.04

15
Find p when b = 48 and r = 130%. _____   48 x 1.3 = 62.4

16
Find p when b = 58 and r = 63%. _____   36.54

17
Read the following question carefully:

What number is 10% of 412?

Is 412 the base (b), rate (r), or the part (p) of
the problem? _____   base (b) = 412

18
Read the following problem carefully:

What number is 47% of 892?

Find the b and r. _____ _____   b = 892, r = 47%

19
To find the number that is 47% of 892 _____
(multiply, divide).   multiply

20
Find the number that is 47% of 892. _____   892 x .47 = 419.24

21
What number is 15% of 88?   b = 88
   b = _____, r = _____, p = _____   r = 15%
                                                           p = 88 x .15 = 13.20

22
9% of 312 is what number?   b = 312
   b = _____, r = _____, p = _____.   r = 9%
                                                           p = 312 x .09
                                                           p = 28.08

23
Find b, p, and r for:
What number is 43% of 56? _____ _____ _____   b = 56
                                                           r = 43%
                                                           p = 56 x .43 = 24.08

24
Find b, p, and r for:
What number is 18% of 72? _____ _____ _____   b = 72
                                                           r = 18%
                                                           p = 72 x .18 = 12.96

25
Find b, p, and r for:
64% of 900 is what number? _____ _____ _____

b = 900
r = 64%
p = 900 x .64 = 576

᠁᠁᠁   ᠁᠁᠁   ᠁᠁᠁   ᠁᠁᠁   ᠁᠁᠁

Post Quiz #6

This quiz reviews the preceding unit. Answers are at the back of the book.
Do not continue until all problems are understood.

Find b, r, and p for:

1
    12% of 46 is _____.

2
    56% of 334 is _____.

3
    _____ is 48% of 700.

4
    _____ is 125% of 140.

5
    6% of 542 is _____.

6
    _____ is 19% of 413.

7
    84% of 894 is _____.

8
    _____ is 9% of 560.

9
    _____ is 143% of 65.

10
    36% of 491 is _____.

UNIT 7: FINDING THE RATE WHEN THE BASE AND PART ARE KNOWN

*******   **Example**   *******   *******   ******

RATE IN A PERCENT PROBLEM

In a percent problem where the base (b) and part (p) are known, the rate (r) is found by dividing p by b.

$$\text{(part)} \div \text{(base)} = \text{(rate)} \qquad r = p \div b \quad \text{or} \quad r = \frac{p}{b}$$

The division is rounded off at two decimal places and the decimal is changed to a percent.

॥॥॥॥॥॥॥   ॥॥॥॥॥॥॥   ॥॥॥॥॥॥॥   ॥॥॥॥॥॥॥   ॥॥॥॥॥॥

1
The decimal .37 can be changed to 37%. Write the decimal .46 as a percent by moving the decimal point two places to the right. _____   46%

2
Write .06 as a percent. _____   6%

3
Write 1.56 as a percent. _____   156%

4
Write .8 as a percent. _____   80%

5
Divide 31 by 94 and round off the answer at two decimal places. _____

$$\begin{array}{r} .329 \doteq .33 \\ 94\overline{)31.000} \end{array}$$

6
Divide 46 by 802 and round off the answer at two decimal places. _____

$$\begin{array}{r} .057 \doteq .06 \\ 802\overline{)46.000} \end{array}$$

7
Divide 65 by 85 and round off the answer at two decimal places. _____

$$\begin{array}{r} .764 \doteq .76 \\ 85\overline{)65.000} \end{array}$$

8
Divide 82 by 48 and round off the answer at two
decimal places. _____

$$\begin{array}{r} 1.708 \\ 48\overline{\smash{)}82.000} \end{array} \doteq 1.71$$

9
Divide 46 by 308, round off at two decimal places,
and write the answer as a percent. _____

$$\begin{array}{r} .149 \\ 308\overline{\smash{)}46.000} \end{array} \doteq 15\%$$

10
A percent problem has three numbers called base,
rate, and part. What is the base? _____

      63 is _____% of 126.

126

11
In the problem below, b = 126. Is 63 the
rate? _____

      63 is _____% of 126

No, p = 63

12
In the problem below, b = 126 and p = 63.

    63 is _____% of 126

Find r by dividing 63 by 126. _____

$$\begin{array}{r} .5 \\ 126\overline{\smash{)}63.0} \end{array} = 50\%$$

13
Find b for:

_____% of 437 is 85. _____

b = 437

14
Find r for:

_____% of 437 is 85. _____

(Divide 85 by 437.)

$$\begin{array}{r} .194 \\ 437\overline{\smash{)}85.000} \end{array} \doteq 19\%$$

15
In percent problems where the base (b) and part
(p) are known the part is divided by the base to
find the rate. Which of the following is the
correct formula for finding the rate? _____

    $r = \dfrac{b}{p}$ or $r = \dfrac{p}{b}$

$r = \dfrac{p}{b}$

16
Find the rate for the following problem by dividing the part by the base. _____

_____% of 812 is 56.

$$\begin{array}{r} .068 \\ 812\overline{)56.000} \end{array} \pm 7\%$$

17
Find the rate for the following problem by dividing part by base. _____

_____% of 76 is 29.

$$\begin{array}{r} .381 \\ 76\overline{)29.000} \end{array} \pm 38\%$$

18
Find the rate for the following problem by dividing part by base. _____

46 is _____% of 258.

$$\begin{array}{r} .178 \\ 258\overline{)46.000} \end{array} \pm 18\%$$

19

_____% of 436 is 89.

$$\begin{array}{r} .204 \\ 436\overline{)89.000} \end{array} \pm 20\%$$

20

_____% of 58 is 96.

$$\begin{array}{r} 1.655 \\ 58\overline{)96.00} \end{array} \pm 166\%$$

21

_____% of 306 is 12.

$$\begin{array}{r} .039 \\ 306\overline{)12.000} \end{array} \pm 4\%$$

22

19 is _____% of 26.

$$\begin{array}{r} .730 \\ 26\overline{)19.000} \end{array} \pm 73\%$$

23

112 is _____% of 75.

$$\begin{array}{r} 1.493 \\ 75\overline{)112.000} \end{array} \pm 149\%$$

24

56 is _____% of 806.

$$\begin{array}{r} .069 \\ 806\overline{)56.000} \end{array} \pm 7\%$$

25

46 is _____% of 50.

$$50 \overline{)46.00} \quad \frac{.92}{} = 92\%$$

26

_____% of 3 is 4.

$$3 \overline{)4.000} \quad \frac{1.333}{} \doteq 133\%$$

Post Quiz #7

This quiz reviews the preceding unit. Answers are at the back of the book. Do not continue until all problems are understood.

Find the b, p, and r for the following:

1
_____% of 438 is 97.

2
_____% of 96 is 8.64.

3
46 is _____% of 64.

4
31 is _____% of 468.

5
_____% of 46 is 76.

6
_____% of 108 is 59.

7
36 is _____% of 248.

8
58 is _____% of 248.

9
_____% of 85 is 33.

10
_____% of 104 is 115.

UNIT 8: FINDING THE BASE WHEN THE RATE AND PART ARE KNOWN

*******   **Example**   *******   *******   ******

THE BASE IN A PERCENT PROBLEM

When the rate (r) and the part (p) of a percent problem are known the base is found by dividing the part by the rate.

$$\text{(part)} \div \text{(rate)} = \text{(base)} \qquad b = p \div r \quad \text{or} \quad b = \frac{p}{r}$$

Before dividing, the rate must be written as a decimal by moving its decimal point two places to the left.

1

The base (b) of a percent statement usually is directly after the word "of". What is the base of the following problem? _____

     15% of 300 is 45.

b = 300

2

In the following problem what is the rate?
     6% of _____ is 2.4   _____

6% = .06

3

To find the base for the following problem the part is divided by the rate.

     6% of _____ is 2.4

Divide 2.4 by .06 _____

$$\overset{40.}{.06 \overline{)2.40}}$$

4

Find the base for the problem at the right by dividing part by rate. _____

8% of _____ is 16

$$\overset{2\ 00.}{.08 \overline{)16.00}}$$

5

Find the base for the problem at the right by dividing part by rate. _____

15% of _____ is 60

$$\overset{4\ 00.}{.15 \overline{)60.00}}$$

6

Find the missing number of:

    25% of _____ is 48

$$\begin{array}{r} 1\ 92. \\ .25.\overline{)48.00} \end{array}$$

7

    37% of _____ is 115.44

$$\begin{array}{r} 3\ 12. \\ .37.\overline{)115.44} \end{array}$$

8

    41% of _____ is 95.12

$$\begin{array}{r} 2\ 32. \\ .41.\overline{)95.12} \end{array}$$

9

    7.8 is 12% of _____

$$\begin{array}{r} 65. \\ .12.\overline{)7.80} \end{array}$$

10

    252.2 is 52% of _____

$$\begin{array}{r} 4\ 85. \\ .52.\overline{)252.20} \end{array}$$

11

    66% of _____ is 18.48

$$\begin{array}{r} 28. \\ .66.\overline{)18.48} \end{array}$$

12

    254.04 is 73% of _____

$$\begin{array}{r} 3\ 48. \\ .73.\overline{)254.04} \end{array}$$

13

    85% of _____ is 566.95

$$\begin{array}{r} 6\ 67. \\ .85.\overline{)566.95} \end{array}$$

14

    110% of _____ is 55

$$\begin{array}{r} 50. \\ 1.10.\overline{)55.00} \end{array}$$

15

    76.72 is 137% of _____

$$\begin{array}{r} 56. \\ 1.37.\overline{)76.72} \end{array}$$

16

    4% of _____ is 10.16

$$\begin{array}{r} 2\ 54. \\ .04.\overline{)10.16} \end{array}$$

17

    28.89 is 9% of _____

$$\begin{array}{r} 3\ 21. \\ .09.\overline{)28.89} \end{array}$$

ıııııııııı   ıııııııııı   ıııııııııı   ıııııııııı   ıııııııı

Post Quiz #8

This quiz reviews the preceding unit. Answers are at the back of the book. Do not continue until all problems are understood.

Find b.

1

    7% of _____ is 21

2

    43% of _____ is 258

3

    35% of _____ is 28

4

    60% of _____ is 90

5

    63 is 45% of _____

6

    5% of _____ is 43.6

7

    18% of _____ is 97.2

8

    214.62 is 49% of _____

9

    275.6 is 130% of _____

10

    16% of _____ is 5.6

UNIT 9: SOLVING PERCENT PROBLEMS .

******   Example   *******   *******   ******

FINDING THE BASE, RATE, OR PART IN A PERCENT PROBLEM

The triangle shown at the right can be used to solve any of the three types of percent problems.

1) If the base (b) and the rate (r) are known, the part (p) is found by the formula

$$p = r \times b$$

2) If the base (b) and the part (p) are known, the rate (r) is found by the formula

$$r = \frac{p}{b}$$

3) If the rate (r) and the part (p) are known, the base (b) is found by the formula

$$b = \frac{p}{r}$$

ⅢⅢⅢⅢ   ⅢⅢⅢⅢ   ⅢⅢⅢⅢ   ⅢⅢⅢⅢ   ⅢⅢⅢ

1
The triangle at the right shows three formulas for solving percent problems. To find the part (p) cover the letter p. The triangle shows that r and b are _____ (multiplied, divided).

multiplied

2
The triangle at the right shows how to find the r when p and b are known. The correct formula is r = _____.

$$r = \frac{p}{b}$$

3
The triangle at the right shows
how to find b when r and p are
known. The correct formula is
b = _____ .

$b = \dfrac{p}{r}$

4
Use the triangle to write the
formula for p.

_____

$p = r \times b$

5
Use the triangle to write a formula for r.

r = _____

$r = \dfrac{p}{b}$

6
If b = 82 and r = 18%, find p. _____

$p = 82 \times .18 = 14.76$

7
If b = 24 and p = 6, find r. _____

$r = \dfrac{6}{24} = 25\%$

8
If r = 6% and p = 9, find b. _____

$b = \dfrac{9}{.06} = 150$

9
If r = 43% and b = 408, find p. _____

$p = .43 \times 408$
$p = 175.44$

10
If b = 56 and p = 21, find r. _____

$r = \dfrac{21}{56} \doteq 38\%$

11
The three numbers in a percent problem need
to be identified. The word "of" often appears
directly before the _____ .

base

211

Chapter 9

12
The three numbers in a percent problem need to be identified. The symbol % is with the _____.

rate

13
    40% of 240 is _____.

$p = 240 \times .40 = 96$

14
    _____ is 15% of 410.

$p = .15 \times 410$
$p = 61.5$

15
    96 is _____% of 512.

$r = \frac{96}{512} \doteq 19\%$

16
    46% of _____ is 51.52.

$b = 112$

17
    19.24 is 52% of _____.

$b = 37$

18
    _____% of 46 is 84.

$r = \frac{84}{46} \doteq 183\%$

19
Fill in the blanks below with letters or numbers to fit the following statement. A class of 46 students has 19 boys.

    _____% of _____ is _____

r% of 46 is 19

20
A class of 46 students has 19 boys. What percent of the class is boys? _____

$b = 46, p = 19$
$r = \frac{19}{46} \doteq 41\%$

21
Fill in the blanks below for the following statement. 75% of a test with 60 questions is true-false.

    _____% of _____ is _____

75% of 60 is _____

212

22

    75% of a test with 60 questions is true-false. How many true-false questions are on the test? _____

b = 60, r = 75%,
p = .75 x 60 = 45

23

An $80 bicycle was reduced $8 for a sale.

    _____% of $80 = $8

    b = _____, p = _____, r = _____

b = 80, p = 8
$r = \frac{8}{80} = \frac{1}{10} = .1$

24

An $80 bicycle was reduced $8 for a sale.

    _____% of $80 = $8

b = 80, p = 8, $r = \frac{1}{10} = .1$

$8 is what percent of the original price? _____

10%

25

    22% of a solution of rubbing alcohol is water. There were 3.96 liters of water in the solution.

    22% of _____ = 3.96

    r = _____, p = _____, b = _____

r = 22%, p = 3.96
b = 3.96 ÷ .22
b = 18 liters

26

Last year Mr. Gottza paid $27,000 for his income taxes which was 38% of his income.

    38% of _____ = $27,000

What was his income? _____

b = $27,000 ÷ .38
b ≐ $71,052.63

27

    2% of the students in a school have a hearing impairment. There are 1500 students in the school.

    2% of 1500 = _____

Find the number of students with a hearing impairment. _____

p = .02 x 1500
p = 30 students

28
Does the question below ask for base, rate, or part? _____

> What percent of a ring weighing 20 grams is silver if there are 8 grams of silver in the ring?

rate

29
Fill in the blanks below for the following question.

> What percent of a ring weighing 20 grams is silver if there are 8 grams of silver in the ring?
> _____% of _____ = _____

r% of 20 = 8

30
Find the percent of silver in a ring of the previous question by finding the r in:

> r% of 20 = 8

$r = \dfrac{8}{20} = 40\%$

31
What is the size of a rock if 3% of its weight is 36 kilograms? Does the previous question ask for the base, rate or part? _____

base

32
What is the size of a rock if 3% of its weight is 36 kilograms? Fill in the blanks below. Place the numbers and the letter b where they belong.

> _____% of _____ = _____

3% of b = 36

33
Find the size of the rock in the previous frame by finding b in:

> 3% of b = 36

b = 36 ÷ .03
b = 1200 kgm

34
How many girls are in a class of 450 students if 52% of the class is girls? Does the previous question ask for the base, rate or part? _____

part

35
Complete the solution of: How many girls are in a class of 450 students if 52% of the class is girls?

52% of 450 = _____   p = r x b

$$p = .52 \times 450$$
$$= 234 \text{ girls}$$

36
Complete the solution of: The interest on a loan of $2000 is $360 for one year. Find the interest rate charged.

r% of $2000 = $360 $r = \dfrac{p}{b} =$ _____

$$r = \frac{360}{2000} = 18\%$$

37
Complete the solution: 30% of the patients in a hospital had heart problems. If 210 patients had heart problems, how many patients are in the hospital?

30% of _____ = 210

$$b = 210 \div .30$$
$$= 700 \text{ patients}$$

38
A store makes a 6% profit on its total sales of $2,000,000. Find the profit by first identifying 6% and $2,000,000 as the base, rate, or part.

_____ _____ _____

$$r = 6\%$$
$$b = \$2,000,000$$
$$p = .06 \times \$2,000,000$$
$$= \$120,000$$

39
87% of milk is water. If a bottle of milk contains 2.61 liters of water, how much milk is in the bottle? Answer the question by first identifying the numbers 2.61 and 87% as the base, rate, or part.

_____ _____ _____

$$r = 87\%$$
$$p = 2.61$$
$$b = 2.61 \div .87$$
$$b = 3 \text{ liters}$$

40
A bottle of medicine contains 12 liters. 2.6% of the contents is active ingredients. How much of the medicine is active ingredients? _____

.312 liters

41

      92% of the body weight of a 160 lb. male is water. How much of the 160 lbs. is water? _____

$$.92 \times 160$$
$$= 147.2 \text{lbs}$$

42

At a sale, slacks were sold for $15 which was 80% of their original price. What was the original price? _____

$$15 \div .80 = \$18.75$$

43

A three-year-old house is worth 125% of its original price of $36,000. How much is the house worth? _____

$$1.25 \times \$36,000$$
$$\$45,000$$

44

About 22,000,000 American workers are union members. If 25% of the total labor force is union, what is the size of the total labor force? _____

$$88,000,000$$

45

A $600 refrigerator was sold for $360 during a sale, a reduction of $240. The reduction in price is what percent of the original price? _____

$$240 \div 600 = 40\%$$

46

The pollution count in a bay increased from 46 parts per million to 54 parts per million. The increase is what percent of the former rate? _____

$$8 \div 46 \doteq 17\%$$

47

A microwave oven uses 35% less power than a conventional oven which uses 800 watts (w.) to cook a roast. How much power would the microwave oven require for the same roast?

_____

$$800 \times .35$$
$$= 280 \text{ w.}$$
$$800 - 280$$
$$= 520 \text{ w.}$$

Post Quiz #9

This quiz reviews the preceding unit. Answers are at the back of the book.
Do not continue until all problems are understood.

Find b, r, and p for problems 1 through 6.

1

58% of 938 is _____.

2

16 is _____% of 298.

3

50.66 is 34% of _____.

4

_____% of 38 is 10.

5

_____ is 6% of 47.

6

155% of _____ is 93.

7
A 6% commission is paid on sales of $40,000. What is the commission?

8
A team won 30 games which was 40% of the games it played. How many
games were played?

9
        18 girls are in a class of 42 students. What percent of the
class is girls?

10
        80% of a 40 member biology class passed the course. How many
passed the course?

11
A sample of uranium ore weighs 90 kilograms. How much of it is
uranium if 2.5% of the ore is uranium?

12
A lawyer's fee for a lawsuit was 35% of the award for the case. The
fee was $10,500. Find the size of the award.

CHAPTER 9 POST-TEST

This test reviews the objectives of the chapter. The student is expected to know how to do **all** of these problems before finishing the chapter. Answers for this test are at the end of the book.

Write the following as both fractions and decimals

1
  2%

2
  110%

3
  55%

4
  9%

Write each of the following as a percent.

5
  $\frac{1}{2}$

6
  0.2

7
  1.01

8
  0.58

9
Which equality at the right is the formula for finding the base in a percent problem?

a) b = p x r

b) b = $\frac{r}{p}$

c) b = $\frac{p}{r}$

d) b = r x p

10
In the statement r% of s = t,

the base is _____

the rate is _____

and the part is _____.

11
  _____% of 80 is 40.

12
  3% of _____ is 9.

13
  10% of 200 is _____.

14
The interest rate on a $50,000 loan for one year is 12%. Find the interest.

15
A 20 liter sample of water contains 1 liter of pollution. What percent of the sample is pollution?

16
An inspector found that 2% of a shipment of meat was spoiled. If 46 kilograms of the meat was spoiled, how much meat was in the shipment?

17
  Find 152% of 46.

18
  10 is 125% of what number?

Chapter 10

What Is the Source of Power, Worth, and Self-Esteem?

The title of this chapter is related to the following quote from Martin Seligman's book **Helplessness.**

> **"From where does one get a sense of power, worth, and self-esteem? Not from what he owns, but from long experiences watching his own actions change the world."** *

The individual who feels she/he has control over the events around her/him derives power, worth, and positive self-esteem from that control. Control in a math learning situation consists of three factors.

*Seligman, Martin, **Helplessness,** W. H. Freeman and Co., San Francisco: 1975, p. 98

Chapter 10

CONTROLLING MATH LEARNING SITUATIONS

To gain control over math learning situations, three factors are of crucial importance. They are:

1) A realistic understanding of the situation.

2) Active, rather than passive, behavior.

3) Repeated success.

These factors have all been designed into this book. This chapter provides another opportunity for experiencing repeated success on mathematical tasks.

ΠΠΠΠΠΠΠΠ   ΠΠΠΠΠΠΠΠ   ΠΠΠΠΠΠΠΠ   ΠΠΠΠΠΠΠΠ   ΠΠΠΠΠΠΠΠ

UNIT 1: THE MEANING OF A RATIO

*******   **Example**   *******   *******   ******

RATIO

A **ratio** is a comparison of two numbers. Three ways to show the same **ratio** are listed below.

$$2 \text{ to } 3 \qquad 2{:}3 \qquad \frac{2}{3}$$

This unit will be concerned with the first and third method of showing a **ratio**.

ΠΠΠΠΠΠΠΠ   ΠΠΠΠΠΠΠΠ   ΠΠΠΠΠΠΠΠ   ΠΠΠΠΠΠΠΠ   ΠΠΠΠΠΠΠΠ

1
The **ratio** of 4 to 5 can be expressed as the fraction $\frac{4}{5}$. 4 is the numerator and 5 is the denominator. Give the **ratio** of 7 to 4 as a fraction. _____

$$\frac{7}{4}$$

2

The **ratio** of 7 to 6 is the fraction with the numerator 7 and the denominator 6.   Give the **ratio** of 7 to 6. _____

$\frac{7}{6}$

3

The ratio of 3 to 11 is the fraction $\frac{3}{11}$.   Which fraction shown below is the ratio of 8 to 7? _____

$\frac{8}{7}$ $\frac{7}{8}$

$\frac{8}{7}$

4

Does $\frac{9}{6}$ show the ratio of 6 to 9? _____

No, 9 to 6

5

Give the ratio of 10 to 32 as a fraction. _____

$\frac{10}{32}$

6

The ratio of 28 to 18 is shown by the fraction _____.

$\frac{28}{18}$

7

The ratio of 2.7 to 0.8 is shown by the fraction _____.

$\frac{2.7}{0.8}$

8

The ratio of 18 to 7.8 is the fraction _____.

$\frac{18}{7.8}$

9

When n and d represent numbers, the ratio of n to d is shown by the fraction $\frac{n}{d}$.
$\frac{6}{9}$ shows the ratio of _____ to _____.

6 to 9

10

$\frac{7}{11}$ shows the ratio of 7 to 11.

$\frac{9}{23}$ shows the ratio of _____ to _____.

9 to 23

11

Does $\frac{0.4}{8}$ show the ratio of 0.4 to 8? _____

yes

12
Does $\frac{1.2}{13}$ show the ratio of 13 to 1.2? _____

no, 1.2 to 13

13
Does $\frac{0.008}{0.01}$ show the ratio of 0.008 to 0.01? _____

yes

14
The ratio of 12 sophomores to 8 freshmen is
$\frac{12}{8}$ which simplifies to $\frac{3}{2}$.
Give the ratio of 4 books to $48 in simplest
terms. _____

$\frac{1}{12}$

15
Give the ratio of 25 drops to 20 liters in
simplest terms. _____

$\frac{5}{4}$

16
Give the ratio of 180 kilometers to 6 hours in
simplest terms. _____

$\frac{180 \div 6}{6 \div 6} = \frac{30}{1}$

17
Give the ratio of 40 units to 80 kilograms in
simplest terms. _____

$\frac{40 \div 40}{80 \div 40} = \frac{1}{2}$

18
Give the ratio of 210 cycles to 60 minutes in
simplest terms. _____

$\frac{210 \div 30}{60 \div 30} = \frac{7}{2}$

19
Give the ratio of $20 to 5 liters of whiskey in
simplest terms. _____

$\frac{4}{1}$

20
Give the ratio of 4,000 particles to 8,000 particles
of pollution in simplest terms. _____

$\frac{1}{2}$

21
Give the simplest form of the ratio of 4,000 millirems of radiation to 6 hours. _____

$$\frac{2,000}{3}$$

22
The ratio $\frac{0.1}{3.2}$ can be simplified by first

multiplying both numerator and denominator by 10. The multiplier 10 is used because one digit is right of the decimal in the two components of the ratio.

$$\frac{0.1}{3.2} = \frac{0.1 \times 10}{3.2 \times 10} = \frac{1.}{32.} = \underline{\quad}$$

$$\frac{1}{32}$$

23
Complete the simplification of the ratio of 2.8 kilometers to 2 seconds. _____

$$\frac{2.8}{2} = \frac{2.8 \times 10}{2 \times 10} = \frac{28}{20} = \underline{\quad}$$

$$\frac{7}{5}$$

24
Simplify the ratio of 3.8 grams to 1.25 seconds.

$$\frac{3.8}{1.25} = \frac{3.8 \times 100}{1.25 \times 100} = \underline{\quad}$$

$$\frac{380}{125} = \frac{76}{25}$$

25
Simplify the ratio of 4 eggs to 1.25 cups of milk.

$$\frac{4}{1.25} = \frac{4 \times 100}{1.25 \times 100} = \underline{\quad}$$

$$\frac{400}{125} = \frac{16}{5}$$

26
Simplify the ratio of 1.362 mg to 4 liters.

$$\frac{1.362}{4} = \frac{1.362 \times 1000}{4 \times 1000} = \underline{\quad}$$

$$\frac{1362}{4000} = \frac{681}{2000}$$

27
Simplify the ratio of 0.3 cm to 5 kilometers.

_____

$$\frac{0.3 \times 10}{5 \times 10} = \frac{3}{50}$$

28
Simplify the ratio of 7.5 kilometers to 0.5 centimeters. _____

$$\frac{7.5 \times 10}{0.5 \times 10} = \frac{15}{1}$$

Chapter 10

29
Simplify the ratio of 4.5 grams to 5.5 milliliters.
_____

$$\frac{4.5 \times 10}{5.5 \times 10} = \frac{9}{11}$$

30
Simplify the ratio of $2.61 to $2. _____

$$\frac{2.61 \times 100}{2 \times 100} = \frac{261}{200}$$

31
Simplify the ratio of 0.76 mg to 1.8 liters. _____

$$\frac{0.76 \times 100}{1.8 \times 100} = \frac{19}{45}$$

᠁᠁᠁᠁᠁   ᠁᠁᠁᠁᠁   ᠁᠁᠁᠁᠁   ᠁᠁᠁᠁᠁   ᠁᠁᠁᠁

Post Quiz #1

This quiz reviews the preceding unit. Answers are at the back of the book.
Do not continue until all problems are understood.

1
Does $\frac{7}{13}$ show the ratio of 7 to 13?

2
Give the ratio of 6 to 13.

3
Give the ratio of 1.3 to 7.

4
Give the ratio of 0.8 to 0.3.

Give the following ratios in simplest form.

5
    8 to 10

6
    240 km to 2 hrs.

7
    18 problems correct to 20
problems total.

8
    $1,260 total price to 4
refrigerators

9
    3.4 grams to 0.5 cu. cc

10
    403 grams to 5.51 liters

11
    0.58 kilograms to 23 cu. cc

12
    4,756 kilometers to 2.3 hrs.

UNIT 2: THE MEANING OF PROPORTIONS

*******   Example   *******   *******   ******

PROPORTIONS

A **proportion** is an equality of two ratios.

Three examples of **proportions** are given below.

$$\frac{2}{3} = \frac{4}{6} \qquad\qquad \frac{6}{4} = \frac{18}{12} \qquad\qquad \frac{50}{100} = \frac{25}{50}$$

1
Two ratios with an equal sign between them form a
proportion as shown below.
$$\frac{8}{10} = \frac{5}{4}$$
Is $\frac{20}{200}$ a proportion? _____   No

2
The statement below is a **proportion.**
$$\frac{60}{100} = \frac{30}{50}$$
Is $\frac{3}{5} = \frac{6}{10}$ a proportion? _____   Yes

3
Is $\frac{5}{8} = \frac{15}{24}$ a proportion? _____   Yes

4
Is $\frac{600}{1000} = \frac{6}{10}$ a proportion? _____   Yes

5
$\frac{?}{8} = \frac{2}{4}$ is a proportion with one term missing.

Is $\frac{15}{15} = \frac{?}{8}$ a proportion? _____   Yes

6
$\frac{n}{8} = \frac{3}{16}$ is a proportion with the letter n
holding the place for a number.
Is $\frac{n}{9} = \frac{2}{31}$ a proportion? _____   Yes

7

Three proportions are shown below.

$$\frac{?}{18} = \frac{2}{3} \qquad \frac{5}{n} = \frac{17}{39} \qquad \frac{37}{50} = \frac{74}{100}$$

Is $\frac{13}{4} = \frac{5}{n}$ a proportion? _____   Yes

8

Is $\frac{n}{25}$ a proportion? _____   No

9

Is $3n = 28 \times 29$ a proportion? _____   No

10

Is $\frac{15}{n} = \frac{27}{13}$ a proportion? _____   Yes

******   **Example**   ******   ******   ******

THE MEANS AND EXTREMES OF A PROPORTION

For the proportion $\frac{3}{8} = \frac{12}{32}$ the first numerator, 3, and the second denominator, 32, are called **means.**

The first denominator, 8, and the second numerator, 12, are called **extremes.**

In a **true proportion,** the product of the means is equal to the product of the extremes.

$\frac{3}{8} = \frac{12}{32}$ is a **true proportion** because 3 x 32 equals 8 x 12.

꜑꜑꜑꜑꜑꜑   ꜑꜑꜑꜑꜑꜑   ꜑꜑꜑꜑꜑꜑   ꜑꜑꜑꜑꜑꜑   ꜑꜑꜑꜑꜑

11

For the proportion $\frac{1}{2} = \frac{2}{4}$ the **means** are

1 and 4, and the **extremes** are 2 and 2. The product of the means is 1 x 4 or 4. The product of the extremes is _____.   2 x 2 = 4

12
For the proportion $\frac{3}{4} = \frac{12}{16}$ the **means** are
3 and 16 and the **extremes** are 4 and 12. Find
the product of the means and the product of the
extremes. _____ _____

48 in both cases

13
In the proportion $\frac{3}{12} = \frac{1}{4}$ the **means** are
3 and 4 and the extremes are _____ and _____.

12, 1

14
What are the means of the proportion below? _____

$$\frac{15}{18} = \frac{5}{6}$$

15, 6

15
Two ratios form a **true** proportion if the product of
the means is equal to the product of the extremes.

Is $\frac{12}{16} = \frac{4}{5}$ a **true** proportion? _____

No, 12 x 5 = 60
16 x 4 = 64

16
Is $\frac{5}{8} = \frac{15}{24}$ a true proportion? _____

Yes, 5 x 24 = 120
8 x 15 = 120

17
Decide if $\frac{7}{14} = \frac{9}{18}$ is a true
proportion by comparing the products of the
means and the extremes. _____

A true proportion,
126 = 126

18
Is $\frac{3}{5} = \frac{4}{6}$ a true proportion? _____

No, 3 x 6 = 18
but 5 x 4 = 20

19
Is $\frac{20}{80} = \frac{2}{8}$ a true proportion? _____

Yes, 20 x 8 = 160
and 80 x 2 = 160

20   Is $\frac{10}{30} = \frac{20}{60}$ a true proportion? _____   Yes,
                                                                              $10 \times 60 = 30 \times 20$

21   Is $\frac{1}{3} = \frac{1}{2}$ a true proportion? _____   No,
                                                                              $1 \times 2 \neq 3 \times 1$

22   Is $\frac{6}{51} = \frac{2}{17}$ a true proportion? _____   Yes,
                                                                              $6 \times 17 = 51 \times 2$

23   Is $\frac{4.5}{9} = \frac{3}{6}$ a true proportion? _____   Yes,
                                                                              $4.5 \times 6 = 9 \times 3$

24   Is $\frac{3}{9} = \frac{6.6}{20}$ a true proportion? _____   No,
                                                                              $3 \times 20 \neq 9 \times 6.6$

25 $\frac{1}{2} = \frac{2}{4}$ is a true proportion because
                        $1 \times 4 = 2 \times 2.$

   $\frac{3}{4} = \frac{6}{8}$ is a true proportion because
                        $3 \times 8 = 4 \times 6.$

   $\frac{9}{12} = \frac{3}{4}$ is a true proportion because
                        $9 \times 4 = $ _____ . 12×3

26 $\frac{2}{7} = \frac{8}{28}$ is a true proportion because
                        $2 \times 28 = 7 \times 8.$

   $\frac{3}{5} = \frac{9}{15}$ is a true proportion because
                        $3 \times 15 = $ _____ . 5×9

27 $\frac{1}{6} = \frac{5}{30}$ is a true proportion because
                        _____ = _____ . $1 \times 30 = 6 \times 5$

28 $\frac{5}{10} = \frac{1}{2}$ is a true proportion because
                        _____ = _____ . $5 \times 2 = 10 \times 1$

29
$\frac{36}{24} = \frac{5}{8}$ is not a true proportion because
_____ ? _____

$36 \times 8 \neq 24 \times 5$

******* **Example** ******* ******* ******

THE USE OF THE LETTER "N" AS A PLACEHOLDER

The letter "n" will be used to hold the place for a number in the proportions of this unit.

When only three of the components of a proportion are known, the letter "n" will represent the fourth component.

Ways to find the value of n's replacement that will produce a true proportion will be shown in the following unit.

30
$\frac{n}{17} = \frac{5}{6}$ is a true proportion
if n x 6 = 17 x 5.

$\frac{n}{8} = \frac{2}{9}$ is a true proportion
if n x 9 = _____ .

8×2

31
$\frac{6}{n} = \frac{13}{14}$ is a true proportion
if 6 x 14 = n x 13.

$\frac{8}{13} = \frac{9}{n}$ is a true proportion
if _____ = _____ .

$8 \times n = 13 \times 9$

32
$\frac{47}{n} = \frac{28}{33}$ is a true proportion
if _____ = _____ .

$47 \times 33 = n \times 28$

33
$\frac{2}{9} = \frac{n}{18}$ is a true proportion
if _____ = _____ .

$2 \times 18 = 9 \times n$

34
$\frac{7}{8} = \frac{14}{n}$ is a true proportion
if _____ = _____ .

$7 \times n = 8 \times 14$

35

$\frac{n}{19} = \frac{23}{12}$ is a true proportion
if _____ = _____ .

$n \times 12 = 19 \times 23$

36

$\frac{4}{n} = \frac{21}{47}$ is a true proportion
if _____ = _____ .

$4 \times 47 = n \times 21$

ⅿⅿⅿⅿⅿ   ⅿⅿⅿⅿⅿ   ⅿⅿⅿⅿⅿ   ⅿⅿⅿⅿⅿ   ⅿⅿⅿⅿ

Post Quiz #2

This quiz reviews the preceding unit. Answers are at the back of the book.
Do not continue until all problems are understood.

In problems 1 through 7 use the products of the means and extremes
to determine if the ratios form a true proportion.

1    $\frac{1}{2} = \frac{10}{20}$

2    $\frac{3}{8} = \frac{15}{40}$

3    $\frac{28}{99} = \frac{2}{7}$

4    $\frac{9}{18} = \frac{13}{26}$

5    $\frac{11}{22} = \frac{15}{32}$

6    $\frac{2.5}{5} = \frac{2.4}{4.8}$

7    $\frac{4}{8} = \frac{0.2}{4}$

8    $\frac{n}{8} = \frac{2}{16}$ if _____ = _____

9    $\frac{37}{18} = \frac{n}{10}$ if _____ = _____

10    $\frac{16}{20} = \frac{5}{n}$ if _____ = _____

11    $\frac{7}{n} = \frac{1}{2}$ if _____ = _____

12    $\frac{3}{n} = \frac{7}{13}$ if _____ = _____

UNIT 3: SOLVING SIMPLE MULTIPLICATION EQUATIONS

*******   Example   *******   *******   ******

SOLVING STATEMENTS OF THE FORM 5 x n = 75

To **solve** 5 x n = 75:

1) Divide both 5 x n and 75 by 5. $\frac{5 \times n}{5} = \frac{75}{5}$

2) Simplify both sides of the equality. $\frac{5 \times n}{5} = n$ and $\frac{75}{5} = 15$

Therefore, n = 15   15 is the solution of 5 x n = 75.

The truth of the previous statement is confirmed by showing that 5 x 15 = 75 is true.

1
To **solve** 2 x n = 24, find the number that is multiplied by 2 to give 24.

Solve 2 x n = 24. _____   12; 2 x 12 = 24

2
To **solve** 20 = 4 x n, find the number that is multiplied by 4 to give 20.

Solve 20 = 4 x n. _____   5; 20 = 4 x 5

3
To solve 3 x n = 18, find the number that is multiplied by 3 to give 18.

Solve 3 x n = 18. _____   6; 3 x 6 = 18

4
Solve n x 5 = 15 by finding the number that is multiplied by 5 to give 15. _____   3; 3 x 5 = 15

5
Solve 4 x n = 24. _____   6; 4 x 6 = 24

6
Solve 38 = 2 x n. _____   19; 38 = 2 x 19

7
Solve 3 x n = 12. _____   4; 3 x 4 = 12

8
Solve n x 2 = 28. _____   14

9
Solve 36 = n x 9. _____   4

10
Solve 8 = 2 x n. _____   4

11
Solve n x 5 = 45. _____   9; 9 x 5 = 45

NEW IDEA

12
To solve 5 x n = 42, it is necessary to find a
number that is multiplied by 5 to give 42.
Is the number a whole number? _____   No

13
The solution to 5 x n = 42 is not a whole number.
To solve 5 x n = 42, divide 42 by 5.

Write $\frac{42}{5}$ as a mixed number. _____ $8\frac{2}{5}$

14
The solution to 5 x n = 42 is $\frac{42}{5}$ or $8\frac{2}{5}$.

Find the answer for 3 x n = 20 by dividing 20
by 3. _____ $\frac{20}{3} = 6\frac{2}{3}$

NEW IDEA

15
Solve 6 x n = 31 by dividing 31 by 6. _____

$$\frac{31}{6} = 5\frac{1}{6}$$

16
Solve 4 x n = 27 by dividing 27 by 4. _____

$$\frac{27}{4} = 6\frac{3}{4}$$

17
Solve 9 x n = 21 by dividing 21 by 9. _____

$$\frac{21}{9} = 2\frac{1}{3}$$

18
Solve 6 x n = 5 by dividing 5 by 6. _____

$$\frac{5}{6}$$

19
Solve 8 x n = 33. _____

$$\frac{33}{8} = 4\frac{1}{8}$$

20
Solve 6 x n = 47. _____

$$\frac{47}{6} = 7\frac{5}{6}$$

21
Solve 2 x n = 19. _____

$$\frac{19}{2} = 9\frac{1}{2}$$

22
Solve 7 x n = 4. _____

$$\frac{4}{7}$$

23
Solve 4 x n = 19. _____

$$\frac{19}{4} = 4\frac{3}{4}$$

24
Solve 8 x n = 44. _____

$$\frac{44}{8} = 5\frac{1}{2}$$

25
Solve 8 x n = 6. _____

$$\frac{6}{8} = \frac{3}{4}$$

26
Solve 6 x n = 50. _____

$$\frac{50}{6} = 8\frac{1}{3}$$

NEW IDEA

27

The equation 14 = 3 x n has 3 multiplied by n.
Solve 14 = 3 x n by dividing 14 by 3. _____

$\frac{14}{3} = 4\frac{2}{3}$

28

Solve 16 = 5 x n by dividing 16 by 5. _____

$\frac{16}{5} = 3\frac{1}{5}$

29

To solve 19 = 5 x n divide 19 by _____.

5

30

Solve 19 = 5 x n. _____

$\frac{19}{5} = 3\frac{4}{5}$

31

To solve 6 x n = 35 divide _____ by _____.

35, 6

32

Solve 6 x n = 35. _____

$\frac{35}{6} = 5\frac{5}{6}$

33

To solve 10 = 15 x n divide _____ by _____.

10, 15

34

Solve 10 = 15 x n. _____

$\frac{10}{15} = \frac{2}{3}$

35

Solve 9 x n = 13. _____

$\frac{13}{9} = 1\frac{4}{9}$

36

Solve 32 = 7 x n. _____

$\frac{32}{7} = 4\frac{4}{7}$

37

Solve 12 x n = 4. _____

$\frac{4}{12} = \frac{1}{3}$

Post Quiz #3

This quiz reviews the preceding unit. Answers are at the back of the book.
Do not continue until all problems are understood.

Which is correct (a or b) in question 1 and 2?

1
To solve 5 x n = 18:
   a) divide 5 by 18
   b) divide 18 by 5

2
To solve 8 = 11 x n:
   a) divide 8 by 11
   b) divide 11 by 8

3
   Solve 12 = 2 x n

4
   Solve 17 x n = 21

5
   Solve 7 x n = 63

6
   Solve 6 x n = 20

7
   Solve 37 = 5 x n

8
   Solve 9 x n = 4

9
   Solve 16 = 7 x n

10
   Solve 8 x n = 28

UNIT 4: SOLVING PROPORTIONS

******* Example ******* ******* ******

THE STEPS IN SOLVING A PROPORTION EQUATION

To solve the proportion $\frac{7}{n} = \frac{9}{10}$

1) Use the products of the means and extremes to write a multiplication equation.

$7 \times 10 = 9 \times n$

2) Solve 7 x 10 = 9 x n or 70 = 9 x n.

$n = \frac{70}{9} = 7\frac{7}{9}$

237

1

To solve the proportion $\frac{n}{5} = \frac{11}{6}$ first multiply the means and the extremes.

$$6 \text{ x } n = \underline{\hspace{1cm}}$$

$5 \text{ x } 11 = 55$

2

$\frac{n}{5} = \frac{11}{6}$ gives $6 \text{ x } n = 55$.

Solve $6 \text{ x } n = 55$.

$$n = \underline{\hspace{1cm}}$$

$\frac{55}{6}$ or $9\frac{1}{6}$

3

To solve $\frac{n}{7} = \frac{4}{9}$ first multiply the means and extremes.

$$9 \text{ x } n = \underline{\hspace{1cm}}$$

$4 \text{ x } 7 = 28$

4

$\frac{n}{7} = \frac{4}{9}$ gives $9 \text{ x } n = 28$.

Solve $9 \text{ x } n = 28$.

$$n = \underline{\hspace{1cm}}$$

$\frac{28}{9} = 3\frac{1}{9}$

5

Write $\frac{8}{n} = \frac{10}{13}$ as a multiplication equation using the products of the means and the extremes.

$$\underline{\hspace{1cm}} = \underline{\hspace{1cm}}$$

$104 = 10 \text{ x } n$

6

$\frac{8}{n} = \frac{10}{13}$ gives $104 = 10 \text{ x } n$.

Solve $104 = 10 \text{ x } n$. $\underline{\hspace{1cm}}$

$\frac{104}{10} = 10\frac{2}{5}$

7

Solve $\frac{7}{11} = \frac{5}{n}$ by first writing a multiplication equation using the products of the means and the extremes. $\underline{\hspace{1cm}}$

$7 \text{ x } n = 55$,
$\frac{55}{7} = 7\frac{6}{7}$

8

Solve $\frac{n}{8} = \frac{3}{5}$ by first writing a multiplication equation using the products of the means and the extremes. $\underline{\hspace{1cm}}$

$5 \text{ x } n = 24$
$\frac{24}{5} = 4\frac{4}{5}$

9
Solve $\frac{n}{2} = \frac{3}{11}$ by first writing a multiplication equation using the products of the means and the extremes. _____

$11 \times n = 6,\ \frac{6}{11}$

10
Solve $\frac{5}{9} = \frac{n}{3}$ by first writing a multiplication equation using the product of the means and the extremes. _____

$15 = 9 \times n$

$\frac{15}{9} = 1\frac{2}{3}$

11
Solve $\frac{6}{11} = \frac{9}{n}$. _____

$6 \times n = 99$

$\frac{99}{6} = 16\frac{1}{2}$

12
Solve $\frac{4}{7} = \frac{1}{n}$. _____

$4 \times n = 7$

$\frac{7}{4} = 1\frac{3}{4}$

13
Solve $\frac{6}{5} = \frac{1}{n}$. _____

$6 \times n = 5,\ \frac{5}{6}$

14
Solve $\frac{4}{3} = \frac{n}{7}$. _____

$28 = 3 \times n$

$\frac{28}{3} = 9\frac{1}{3}$

15
Solve $\frac{12}{5} = \frac{2}{n}$. _____

$12 \times n = 10$

$\frac{10}{12} = \frac{5}{6}$

Chapter 10

This quiz reviews the preceding unit. Answers are at the back of the book.
Do not continue until all problems are understood.

Solve:

1
$$\frac{5}{n} = \frac{8}{10}$$

2
$$\frac{9}{14} = \frac{n}{7}$$

3
$$\frac{6}{7} = \frac{5}{n}$$

4
$$\frac{3}{n} = \frac{11}{2}$$

5
$$\frac{n}{7} = \frac{3}{5}$$

6
$$\frac{n}{8} = \frac{1}{15}$$

7
$$\frac{5}{n} = \frac{2}{7}$$

8
$$\frac{6}{11} = \frac{2}{n}$$

ɪɪɪɪɪɪɪɪɪɪ   ɪɪɪɪɪɪɪɪɪɪ   ɪɪɪɪɪɪɪɪɪɪ   ɪɪɪɪɪɪɪɪɪɪ   ɪɪɪɪɪɪɪɪ

UNIT 5: SOLVING WORD PROBLEMS WITH PROPORTIONS

******* **Example** ******* ******* ******

STEPS IN SOLVING WORD PROBLEMS WITH PROPORTIONS

In this section word problems are solved using proportions. Each
problem involves a three-step approach.

1) Write two ratios showing similar comparisons.

2) Write a proportion making the two ratios equal.

3) Solve the proportion.

ɪɪɪɪɪɪɪɪɪɪ   ɪɪɪɪɪɪɪɪɪɪ   ɪɪɪɪɪɪɪɪɪɪ   ɪɪɪɪɪɪɪɪɪɪ   ɪɪɪɪɪɪɪɪ

1
If the ratio of boys to girls in the sophomore class
is 6 to 5, the fraction way of showing this
comparison is $\frac{6}{5}$. If n represents the number of
boys and 45 is the number of girls, write a fraction
ratio comparing the number of boys to girls. _____

$$\frac{n}{45}$$

240

2

If the ratio of boys to girls in the sophomore class is $\frac{6}{5}$ and there are 45 girls in the class the proportion $\frac{6}{5} = \frac{n}{45}$ shows the comparison of boys to girls. The letter n represents the number of _____ (boys, girls) in the sophomore class.

boys.

3

Solve $\frac{6}{5} = \frac{n}{45}$ by first writing the equality that uses the products of the means and the extremes.

_____

$270 = 5 \times n$

$\frac{270}{5} = 54$

4

The following problem is solved using a three-step approach:

> One pound of hamburger makes spaghetti for three people. How much hamburger is needed to make spaghetti for eight people?

> 1) The ratio $\frac{1}{3}$ compares hamburger amount to people.

If n represents the hamburger needed for eight people, what is the ratio? _____

$\frac{n}{8}$

> 2) The two ratios comparing hamburger amounts to people are $\frac{1}{3}$ and $\frac{n}{8}$.

Write a proportion using the ratios. _____

$\frac{1}{3} = \frac{n}{8}$

> 3) Solve the proportion $\frac{1}{3} = \frac{n}{8}$ and the solution will be the amount of hamburger, in pounds, needed for eight people. _____

$n = \frac{8}{3} = 2\frac{2}{3}$

5

If 2 inches on a map represent 15 miles, how many miles are represented by 11 inches?

1) Write two ratios comparing inches to miles. _____ _____

$\frac{2}{15}$ and $\frac{11}{n}$

2) Write a proportion. _____

$\frac{2}{15} = \frac{11}{n}$

3) Solve the proportion. _____

$\frac{165}{2} = 82\frac{1}{2}$

6

If 4 inches of insulation cost \$2, what is the cost of 10 inches of insulation?

1) Write two ratios comparing inches of insulation to cost. _____ _____

$\frac{4}{2}$ and $\frac{10}{n}$

2) Write a proportion. _____

$\frac{4}{2} = \frac{10}{n}$

3) Solve the proportion. _____

\$5

7

If 6 yards of cloth cost \$9, how much does 20 yards cost?

1) Write two ratios comparing yards of cloth to dollars. _____ _____

$\frac{6}{9}$ and $\frac{20}{n}$

2) Write a proportion. _____

$\frac{6}{9} = \frac{20}{n}$

3) Solve the proportion. _____

\$30

8

If 3 long distance runners cover 51 miles, how many runners would be needed to cover 85 miles?

1) Write two ratios comparing runners to miles. _____ _____

$\frac{3}{51}$ and $\frac{n}{85}$

2) Write a proportion. _____

$\frac{3}{51} = \frac{n}{85}$

3) Solve the proportion. _____

5 runners

9

If a mixture of nuts contains 7 pounds of pecans for every 4 pounds of cashews, how many pounds of pecans should be used with 11 pounds of cashews?

1) Write two ratios comparing pecans to cashews. _____ _____

$\frac{7}{4}$ and $\frac{n}{11}$

2) Write a proportion. _____

$\frac{7}{4} = \frac{n}{11}$

3) Solve the proportion. _____

$19\frac{1}{4}$ pounds

10

A solution of 3 parts medicine to 20 parts alcohol is to be made. The total amount of alcohol will be 240 milliliters (ml). How much medicine is needed?

1) Write the proportion comparing medicine to alcohol. _____

$\frac{3}{20} = \frac{n}{240}$

2) Solve the proportion. _____

36 ml

11

The ratio of profit to cost for a product is 1 to 9. If the profit is $16,200 what is the cost?

1) Write a proportion comparing profit to cost. _____

$\frac{1}{9} = \frac{16200}{n}$

2) Solve the proportion. _____

$145,800

12

5 grains of a drug are mixed with 160 milliliters (ml) of a solution. In a sample containing 2 grains of the drug, how much solution will be used?

1) Write a proportion comparing grains to solution. _____

$\frac{5}{160} = \frac{2}{n}$

2) Solve the proportion. _____

64 ml

13

A recipe calls for 1 cup of sugar with 3 cups of flour. How much sugar is needed with 12 cups of flour?

    1) Write a proportion comparing sugar to flour. _____

$$\frac{1}{3} = \frac{n}{12}$$

    2) Solve the proportion. _____

4 cups sugar

14

An intravenous solution needs 2 liters (l) of glucose mixed with 17 units of blood. How much glucose is needed for 35 units of blood?

    1) Write a proportion comparing glucose to blood. _____

$$\frac{2}{17} = \frac{n}{35}$$

    2) Solve the proportion. _____

$4\frac{2}{17}$ ml

15

A doctor's order calls for 8 drams of drug to be used with 100 milliliters of water. How much water will be needed for 18 drams of drug?

    1) Write a proportion comparing drug to water. _____

$$\frac{8}{100} = \frac{18}{n}$$

    2) Solve the proportion. _____

225 ml

16

    2 centimeters (cm) on a map represent 10 kilometers (km). How many centimeters are needed to show 37 kilometers?

    1) Write a proportion comparing centimeters to kilometers. _____

$$\frac{2}{10} = \frac{n}{37}$$

    2) Solve the proportion. _____

$7\frac{2}{5}$ cm

17

A car used 13 liters (l) of fuel traveling 340 kilometers (km). How many liters of fuel will be used traveling 85 kilometers?

    1) Write a proportion comparing fuel to kilometers. _____

$$\frac{13}{340} = \frac{n}{85}$$

    2) Solve the proportion. _____

$$\frac{13}{4} = 3\frac{1}{4} \text{ liters}$$

18

If 20 liters (l) of paint will cover 210 square meters of wall, how many square meters will 15 liters of paint cover?

    1) Write a proportion comparing paint to wall area. _____

$$\frac{20}{210} = \frac{15}{n}$$

    2) Solve the proportion. _____

$$157\frac{1}{2} \text{ sq. m}$$

19

A store decided to give a discount of 12 cents for every $1 of price. A refrigerator was priced at $300. How much is the discount?

    1) Write a proportion comparing discount to price. _____

$$\frac{12}{1} = \frac{n}{300}$$

    2) Solve the proportion. _____

3600 cents = $36

20

If 1500 20-year-olds die each year per every 1,000,000 20-year-olds, how many will die out of 2,000 20-year-olds?

    1) Write a proportion comparing those that will die with the total. _____

$$\frac{1500}{1000000} = \frac{n}{2,000}$$

    2) Solve the proportion. _____

3 people

21
If 2 nuclear accidents can occur every 360 days,
how many may occur in 900 days?

1) Write a proportion comparing accidents
to days. _____

$$\frac{2}{360} = \frac{n}{900}$$

2) Solve the proportion. _____

5 accidents

ᴜᴜᴜᴜᴜᴜᴜ   ᴜᴜᴜᴜᴜᴜᴜ   ᴜᴜᴜᴜᴜᴜᴜ   ᴜᴜᴜᴜᴜᴜᴜ   ᴜᴜᴜᴜᴜᴜ

Post Quiz #5

This quiz reviews the preceding unit. Answers are at the back of the book.
Do not continue until all problems are understood.

For each problem write a proportion and then solve it:

1
Ace Packing Co. makes 3 cents profit
for every $200 in sales. What is
the profit on a $1000 sale?

2
A nurse is ordered to make a mixture
containing 7 tablespoons of drug for
each 10 liters of solvent. If 2 liters
of solvent are used, how much drug is
needed?

3
A student gets 5 out of every 6
problems correct on a test.   If
there are 180 total problems on the
test, how many did the student
answer correctly?

4
A business found that there was 2 cents
advertising cost for every 9 cents in
sales. On an item selling for $18 what
was the advertising cost?

5
A photograph was 3 inches high and
5 inches wide. If it is enlarged
to be 7 inches high, how wide will
it be?

6
In a 250 kilogram block of alloy, 2
grams out of every 5 grams of alloy are
gold. How much gold is in the block?

7
The National Bank charges interest
of 13 cents for every $1 loaned for
one year. How much interest will
be charged for a loan of $150 for
one year?

8
The ratio of the width to length of a
book is to be 8 to 11.   How wide
should a page be if it is 16 cm long?

CHAPTER 10 POST-TEST

This test reviews the objectives of the chapter. The student is expected to know how to do **all** of these problems before finishing the chapter. Answers for this test are at the end of the book.

Give the following ratios in the simplest terms:

1
   3 to 12

2
   28 to 14

3
   20 to 25

4
   32 kilometers to 28 kilometers

Solve the proportions.

5
$$\frac{n}{8} = \frac{8}{16}$$

6
$$\frac{9}{n} = \frac{3}{5}$$

7
$$\frac{18}{20} = \frac{27}{n}$$

8
$$\frac{12}{5} = \frac{n}{20}$$

9
If the F.C.C. ruled that the ratio of advertising time to program time is to be 2 to 5, how much program time is needed to air 16 minutes of advertising?

10
A weather forecaster found that 3 rainy days occur for every 8 cloudy days. In 20 cloudy days, about how many rainy days should occur?

11
It is found that 2 malfunctions occur for every 45 days of operating a nuclear plant. How many days may pass for 3 malfunctions to occur?

12
A bank charges $11 interest on a $100 loan for one year. How much could be borrowed for a year for $220 interest?

13
A 2 liter sample of river water contained 13 particles of pollution. How many particles would be in 50 liters of the water?

14
3 drams of a drug is to be given for 12 kilograms of body weight. Find the size of a person that would be given 14 drams of the drug.

ANSWERS

Answers Chapter 7

Post Quiz #1
1. 2752   2. 10644   3. 3357   4. 15679   5. sum, total
6. 268 lbs.   7. $56,488   8. different

Post Quiz #2
1. 17   2. 12   3. 11   4. a. 2163 + 3451 = 5614
b. 15487 + 25449 = 40936   c. 2847 + 3302 = 6149
d. 19306 + 30694 = 50000   5. $58   6. 133

Post Quiz #3
1. factors   2. a. 560   b. 56780   c. 6100   d. 4800
3. 8208   4. 442

Post Quiz #4
1. 125 R18   2. 497 R44   3. 214 R12   4. 522 R16   5. 337 R7
6. 111 R2   7. 198 R58   8. 264 R94   9. 609 R10   10. 88 R60
11. 1885 R24   12. 185 R38

Post Quiz #5
1. 46, 912, 19, 38   2. 507 x 18 + 17 = 9143   3. 51   4. $357
5. 1563 cards

CHAPTER 7 POST-TEST
1. 1288   2. 1389   3. 1093   4. $386   5. 252 sophomores
6. 456 orders   7. 291 miles   8. 563 acres   9. 612   10. $2,220
11. factors   12. product   13. 672   14. $3,060
15. 47, 2726, 58   16. 204 R36   17. 65   18. 32 oz.
19. 46, 2990, 65   20. 903 R3   21. 887 R4   22. 243 R2
23. 1210 R3   24. 186 R5   25. 987 R22   26. 1036 R46
27. 1574 R34   28. 60   29. 40

Answers Chapter 8

Post Quiz #1
1. difference   2. sum   3. $4\frac{3}{8}$ 4. $3\frac{5}{12}$ yds   5. $20\frac{5}{8}$
6. $9\frac{4}{5}$ kgm   7. $1\frac{1}{2}$ gm   8. $3\frac{87}{100}$ m

Post Quiz #2
1. $1\frac{3}{7}$ 2. $\frac{1}{2}$ 3. 12   4. 18   5. 16   6. 12
7. 100 coins   8. 105 blocks   9. $4\frac{1}{5}$ tons   10. $14\frac{38}{43}$ = 14 shares

Post Quiz #3
1. 65.108   2. 3609.71   3. 8.0842%   4. 25.388   5. 15.736
6. $0.54745

Post Quiz #4
1. 1.11   2. 0.0037   3. 0.37   4. 53.37   5. 1.83   6. 0.38668
7. 0.0007   8. 11.5   9. 0.5538   10. 22.44   11. 8.7 gm   12. 2.183

Post Quiz #5
1. $237.50   2. 62.4 kgm   3. $4.50   4. $2.88   5. 0.00225 kgm
6. 2.4   7. 800   8. 1   9. $2.26   10. 174   11. $3300

Post Quiz #6
1. 904 tickets   2. 1.94 m   3. $2.57   4. $8.40

CHAPTER 8 POST-TEST

1. $4\frac{1}{4}$ kgm   2. $4\frac{3}{4}$ m   3. $1\frac{3}{4}$ cups
4. $\frac{9}{8}, \frac{2}{3}, \frac{3}{4}$ 5. $2\frac{5}{8}$ 6. $1\frac{4}{5}$ million   7. 66   8. $2,800
9. $\frac{6}{5}, \frac{2}{3}, \frac{5}{9}$ 10. 20   11. $1\frac{5}{6}$ 12. 19   13. $1,500
14. 1, 4, 3   15. 67 + 294 = 361   16. 6.68   17. 0.608
18. 7.18   19. 7.4   20. $149.90   21. 0.0058   22. A, B, C
23. 190 + 188 = 378   24. $1101.75   25. 6.917 kgm   26. 3.98
27. 6.72%   28. 20.72   29. 261.8 mg   30. $5450   31. 22.2
32. 43   33. $78.98   34. 25,800 bushels   35. 143 mgm
36. $3450   37. .48 gallons   38. 94   39. $46.37

Answers Chapter 9

Post Quiz #1

1. a. $\dfrac{17}{100}$ b. $\dfrac{49}{100}$ c. $\dfrac{57}{100}$ d. $\dfrac{143}{100}$ e. $\dfrac{7}{100}$

2. one-hundredths   3. a. 19%   b. 81%   c. 3%   d. 181%   e. 67%

Post Quiz #2

1. 0.27   2. 0.85   3. 0.56   4. 0.04   5. 0.37   6. 1.63
7. 0.92   8. 0.058   9. 0.16   10. 0.473

Post Quiz #3

1. 26%   2. 89%   3. 8%   4. 63%   5. 240%   6. 52%
7. 145%   8. 22%   9. 4.3%   10. 141%   11. 1%   12. 2.5%

Post Quiz #4

1. 38%   2. 56%   3. 80%   4. 33%   5. 30%   6. 52%
7. 48%   8. 44%   9. 119%   10. 4%

Post Quiz #5

1. base, rate, part   2. of   3. %   4. a. 500, 53%, 265
b. 20, 85%, 17   c. 60, 125%, 75   d. 180, 80%, 144

Post Quiz #6

1. 46, 12%, 5.52   2. 334, 56%, 187.04   3. 700, 48%, 336
4. 140, 125%, 175   5. 542, 6%, 32.52   6. 413, 19%, 78.47
7. 894, 84%, 750.96   8. 560, 9%, 50.4   9. 65, 143%, 92.95
10. 491, 36%, 176.76

Post Quiz #7

1. 438, 97, 22%   2. 96, 8.64, 9%   3. 64, 46, 72%   4. 468, 31, 7%
5. 46, 76, 165%   6. 108, 59, 55%   7. 248, 36, 15%   8. 248, 58, 23%
9. 85, 33, 39%   10. 104, 115, 111%

Post Quiz #8

1. 300   2. 600   3. 80   4. 150   5. 140   6. 872
7. 540   8. 438   9. 212   10. 35

Post Quiz #9

1. 938, 58%, 544.04   2. 298, 5%, 16   3. 149, 34%, 50.66   4. 38, 26%, 10
5. 47, 6%, 2.82   6. 60, 155%, 93   7. $2400   8. 75 games   9. 43%
10. 32   11. 2.25 kgm   12. $30,000

CHAPTER 9 POST-TEST

1. $\dfrac{2}{100}$, .02   2. $\dfrac{110}{100}$, 1.10   3. $\dfrac{55}{100}$, .55   4. $\dfrac{9}{100}$, .09

5. 50%   6. 20%   7. 101%   8. 58%   9. c   10. s, r, t
11. 50%   12. 300   13. 20   14. $6,000   15. 5%
16. 2300 kgm   17. 69.92   18. 8

Answers Chapter 10

Post Quiz #1
1. yes
2. $\frac{6}{13}$
3. $\frac{13}{70}$
4. $\frac{8}{3}$
5. $\frac{4}{5}$
6. $\frac{120}{1}$
7. $\frac{9}{10}$
8. $\frac{\$315}{1}$
9. $\frac{34}{5}$
10. $\frac{40300}{551}$
11. $\frac{29}{1150}$
12. $\frac{47560}{23}$

Post Quiz #2
1. yes
2. yes
3. no
4. yes
5. no
6. yes
7. no
8. n x 16 = 8 x 2
9. 37 x 10 = 18 x n
10. 16 x n = 20 x 5
11. 7 x 2 = n x 1
12. 3 x 13 = n x 7

Post Quiz #3
1. b
2. a
3. 6
4. $1\frac{4}{17}$
5. 9
6. $3\frac{1}{3}$
7. $7\frac{2}{5}$
8. $\frac{4}{9}$
9. $2\frac{2}{7}$
10. $3\frac{1}{2}$

Post Quiz #4
1. $6\frac{1}{4}$
2. $4\frac{1}{2}$
3. $5\frac{5}{6}$
4. $\frac{6}{11}$
5. $4\frac{1}{5}$
6. $\frac{8}{15}$
7. $17\frac{1}{2}$
8. $3\frac{2}{3}$

Post Quiz #5
1. 15 cents
2. $1\frac{2}{5}$ tblspns
3. 150 prob.
4. $4
5. $11\frac{2}{3}$ in.
6. 100 kgm
7. $19.50
8. $11\frac{7}{11}$ cm

CHAPTER 10 POST-TEST
1. $\frac{1}{4}$
2. $\frac{2}{1}$
3. $\frac{4}{5}$
4. $\frac{8}{7}$
5. 4
6. 15
7. 30
8. 48
9. 40 min.
10. $7\frac{1}{2}$ days
11. 67.5 days
12. $2000
13. 325 particles
14. 56 kgm